Ridgewood Public Library
125 N. Maple Avenue
Ridgewood, NJ 07450

CARMELO G. MALACRINO

Constructing the Ancient World
Architectural Techniques of the Greeks and Romans

Translated by Jay Hyams

The J. Paul Getty Museum • Los Angeles

Italian edition © 2010 Arsenale Editrice
Carmelo G. Malacrino, Photography
Andrea Darra, Editorial Coordinator
Fabrizio Tolu, Editorial direction
Linda Bovi, Layouts
Giancarlo Malagutti, Drawings (Except as noted below)
Carmelo G. Malacrino, Drawings, pp. 46, 50, 51 (below), 53–54,
78 (above), 79, 81, 84, 89 (below), 91, 94 (above, right), 110, 162 (below)

English translation © 2010 J. Paul Getty Trust

First published in English in 2010 by the J. Paul Getty Museum, Los Angeles

Getty Publications
1200 Getty Center Drive, Suite 500
Los Angeles, California 90049-1682
www.gettypublications.org

Pamela Heath, *Production Coordinator*
Jay Hyams, *Translator*
Mary Cason, *Editor*
Michael Shaw, *Graphic Design and Typesetting*
Kurt Hauser, *Jacket and Binding Design*
Translation, copyediting, and typesetting coordinated by LibriSource Inc.

Printed in Italy

Library of Congress Cataloging-in-Publication Data

Malacrino, Carmelo G.
  [Ingegneria dei Greci e dei Romani. English]
  Constructing the ancient world : architectural techniques of the Greeks and Romans / Carmelo G. Malacrino.
     p. cm.
  Includes bibliographical references.
  First published in Italy in 2009 by Arsenale-Editrice, Verona.
  ISBN 978-1-60606-016-2 (hardcover)
 1. Building—Greece—History. 2. Architecture, Greek. 3. Building—Rome.
 4. Architecture, Roman. I. Title.
 TH16.M3513 2010
 690.0938—dc22
                                                    2009045050

Cover. Remains of the Fountain of Glauke (foreground) and the Temple of Apollo (background), Corinth, Greece. See page 8.

Contents

4 *Introduction*

7 *Natural Building Materials: Stone and Marble*

41 *Clay and Terracotta*

61 *Lime, Mortar, and Plaster*

77 *Construction Techniques in the Greek World*

111 *Construction Techniques in the Roman World*

139 *Engineering and Techniques at the Work Site*

155 *Ancient Hydraulics: Between Technology and Engineering*

175 *Heating Systems and Baths*

187 *Roads, Bridges, and Tunnels*

208 *Glossary*
210 *Bibliography*
213 *Index*

Introduction

Engineering and construction methods, architecture and technical expertise. These were the fields of knowledge that comprised the primary skills incarnated in the figure of the architect in the ancient world. In the first century A.D., Vitruvius in his treatise *De Re Edificatoria (On Architecture)* wrote of architecture as a science, one that depended "upon many disciplines and various apprenticeships which are carried out in other arts" (1.1.1). The activities of an architect were distinguished by craftsmanship, a "continued and familiar practice, which is carried out by the hands in such material as is necessary for the purpose of a design," and technology, which "sets forth and explains things wrought in accordance with technical skill and method." The architect also must be literate and well educated (1.1.4–10):

> An architect must be a man of letters that he may keep a record of useful precedents. By his skill in draftsmanship he will find it easy by colored drawings to represent the effect desired. Mathematics again furnishes many resources to architecture. It teaches the use of rule and compass and thus facilitates the laying out of buildings on their sites. . . . By optics, in buildings, lighting is duly drawn from certain aspects of the sky. By arithmetic, the cost of building is summed up; the methods of mensuration are indicated. . . . Architects ought to be familiar with history because in their works they often design many ornaments about which they ought to render an account to inquirers. . . . Philosophy . . . makes the architect high-minded, so that he should not be arrogant but rather urbane, fair-minded, loyal, and what is most important, without avarice. . . . A man must know music that he may have acquired the acoustic and mathematical relations. . . . He must know the art of medicine in its relation to the regions of the earth . . . and to the characters of the atmosphere.

This book is dedicated to these and other aspects of the ancient *ars edificatoria,* an art that called upon various disciplines and applications, and involved not only the architect and other members of the work site but all of society. From the city officials who commissioned a great public work to the priests in need of a suitable structure for the performance of rituals in a sacred area to the common people who made homes in which to live and perform their daily activities, construction involved every social class. It is no accident that architectural remains are the most important archaeological sources handed down to us by antiquity.

While traveling one often overhears exclamations of wonder from visitors who are impressed by the monumentality of Greek and Roman structures, along with incredulous queries as to how such works could have been achieved without the kinds of machines we have today. Such observations were in part responsible for my decision to write a book on those very themes, in a text accessible to the wider public, especially nonspecialists. Straightforward language and clear definitions of both ancient and modern terminology are accompanied by a selection of illustrations that, it is hoped, will engage the reader in the subject, and show not only details of the technical methods discussed but also the broader background for the works examined. Special attention has been given to ancient sources (literary, epigraphic, and iconographic), as they represent the irreplaceable testimony handed down to us by direct participants in the construction of ancient architecture.

The book begins with a section on construction materials, specifically the stones and marbles used in the Greek and Roman worlds. Their identification, uses in architecture, and quarries, as well as techniques of working them, are presented within their geographical and historical context, including the economic and social systems that shaped the use of and commerce in stone materials. Clay, one of the first raw materials used by humans, is among the most versatile offered by nature, and was used in architecture both as the principal component in "unbaked" mixtures and as the base for products fired in kilns. The subsequent use of lime, mortar, and plaster led to the creation of structures that are still standing after more than two millennia.

The theme of construction technology occupies the central part of the book. For the sake of simplicity it is divided into separate chapters on the Greek and the Roman worlds, although references to comparisons, parallels, and influences appear throughout.

The final chapters are dedicated to the more complex technical aspects of construction, including the transportation and installation of architectural elements, as well as heating systems, roads, bridges, and tunnels.

The reader will soon discover that, given the title of the book, the discussion could easily have been expanded to include many other subjects, such as pumping systems and methods applied in the field of military engineering. But the decision was made to concentrate on the techniques of ancient building and engineering that were most closely related to the sphere of architecture, and to provide a panorama of examples, at the same time broad and detailed. As the topics considered are certain to arouse interest in further information, a bibliography, divided by more narrowly defined subject areas, has been included. Indeed, the purpose of this book is to introduce the reader to themes and issues that, despite their intrinsic fascination, have largely remained the privileged domain of specialists. These subjects, however, relate to some of the most ingenious solutions devised by the ancient world to resolve the very problems that today we are able to confront only with the help of industrial machines.

NATURAL BUILDING MATERIALS: STONE AND MARBLE

During the passage from the eighth to the seventh century B.C., the entire Greek world was at the center of profound social and economic transformations, most notable among them the evolution of a new form of political organization, the city-state (Greek *polis*). In that same period architecture developed a completely new language, one whose lexicon of forms reflected a great building revolution: the adoption of nonperishable construction materials for the external faces of buildings. This architectural development was a highly complex phenomenon, but many aspects of it are now understood more clearly thanks to recent discoveries and a renewed interest in archaeological research.

In reality, stone had been used in the Greek world for many centuries, to build the great architecture of the Minoan and Mycenaean civilizations. But with the end of those cultures, and the accompanying breakup of the social and economic system embodied by their palaces, the skills and technologies applied to these buildings were lost. The Greeks of the late Geometric period inherited from preceding generations a construction tradition based on modestly sized houses made with natural materials, including clay, wood, and other plant materials. The monumental fortifications of the Mycenaean period—made with blocks that in some cases were more than 4 meters long—thus assumed a mythical aura, as reflected by the Greek traveler and geographer Pausanias (*Description of Greece* 2.16.5) in his remarks on Mycenaean cities: "There still remain, however, parts of the city wall, including the gate, upon which stand lions. These, too, are said to be the work of the Cyclopes, who made for Proteus the wall at Tiryns."

In Italy stone was similarly adopted for architecture over the course of the seventh century B.C. Early in the Orientalizing period, the Etruscans began to use stone as a building material. This process is reflected in several structures at Pian della Civita in Tarquinia made with well-cut blocks that form a framework. A building uncovered on the elevation of the Piazza d'Armi at Veii demonstrates a reliance on stone at Etruscan work sites as early as the second half of the seventh century B.C. The rectangular construction (8.07 by 15.35 meters) is made in blocks of ash-gray tufa with yellowish tones. In Rome the use of stone took hold over the course of the sixth century B.C. with the great architectural program of the Tarquins. A fine example is the impressive stone wall (*murus lapideus*) in blocks of the local stone called cappellaccio, whose construction, according to Livy (*History of Rome* 1.36.1), began under Tarquinius Priscus.

STONE IN GREECE AND THE GREEK WORLD: LIMESTONE, SANDSTONE, AND TUFA

Continental Greece has always been rich in stone materials that are excellent for use in

Opposite. View of the remains of Mycenae, with layers of structures from fortifications and terraced walls.

From earliest times builders at Corinth made use of a soft limestone quarried near the city. Some monuments, such as the Fountain of Glauke in the foreground, were cut directly out of the natural rock.

architecture; the same is true for the Aegean islands, Asia Minor, and various regions colonized by Greek peoples in the West. Of course, the same kinds of stone were not found in every area. Each offered its own peculiar geological deposits, and the adoption of building materials and techniques for their use often resulted in a strongly regional character for architecture. At every construction site efforts were made to use stone that was locally available, with recourse to more-distant quarries only when the material they offered presented superior aesthetic qualities, greater strength, or a higher grade of workability. However, a desire for higher quality often justified the additional costs of transportation—by land, river, or sea—for blocks purchased from a quarry, but also made a significant impact on total construction expenses.

As noted above, the stone (Greek *lithos*, Latin *lapis*) used in ancient construction is of various types. The most widely available were limestone (Gr. *lithos petrinos*, Lat. *lapis albus*), calcarenite, and sandstone, all of sedimentary origin and relatively soft and easy to work. Various types of tufa and porous (Gr. *poros*) limestone were to be found throughout Greece (in Attica and Peloponnesus, and on islands such as Crete and Rhodes) and on the Italian peninsula (in central Italy, Magna Graecia, and Sicily). These stones appeared in a variety of colors and offered good static capacities as well as significant hardness, and were easily worked and dressed with a chisel. As a result, tufa and porous limestone were often adopted for constructing the walls of buildings, as well as for various decorative elements, in the absence of marble or as an alternative to it.

In Athens the first material used on a large scale in stone architecture was the so-called Acropolis limestone, a calcareous breccia that in fact was quarried not only on the upper part of the Acropolis but also on other hills of the city (including the Areopagus, Pnyx, and Lykabettos). This attractive stone features a gray-blue color that sometimes shades toward red, and often exhibits yellowish or brownish red calcite veining. Acropolis limestone was available near construction work sites, if not actually on the sites themselves, and offered the primary advantage of avoiding problems related to long-distance transportation. Nevertheless, as early as the sixth century B.C. Athenians began importing two different types of limestone, both of them softer and more porous than Acropolis limestone, from nearby areas. *Argryleikos lithos*, yellow-pink in color with a fine grain that made possible more detailed work, was quarried in the district of Kara at the foot of Mount Hymettus, about 3.5 kilometers southeast of the city; *aktites lithos*, the softer of the two types and easier to cut, came from Piraeus on the Akti peninsula. During the fifth century B.C. a third third type of soft limestone, *aiginaios lithos*, came into use; it was found in quarries located at an even greater distance from Athens, along the northern coast of the island of Aegina.

Other, more compact types of limestone were extracted in various regions of the Peloponnesus—including the quarry in Argolis used by the Mycenaeans—for construction both at Tiryns and at Mycenae itself; in the fifth century B.C. these materials were also used in Arcadia for the construction of the Temple of Apollo Epikourios at Bassae. At Corinth an oolitic limestone, soft and fine grained, was used in the seventh century B.C. for construction of the first large temple to Apollo; like the Acropolis limestone in Athens, it was quarried from the same bedrock on which the city itself was then being built. Likewise, at Olympia the two main temples and some *thesauroi*, or treasuries, of the sanctuary were erected using the local biocalcarenite, a sedimentary stone full of fossils that was quarried along the banks of the Alpheus. Pausanias (*Desc. of Greece* 5.10.3) described the cult building erected in honor of Zeus as a temple built of native stone in the Doric style, with columns on the exterior.

Many other cities of Magna Graecia had the good fortune to arise in areas with abundant, good-quality building stone. For example, Poseidonia (later the Roman Paestum) had access to a fine local travertine, whose

The frieze of the Erechtheum presented an elegant dichromatic scheme, based on the use of gray-blue limestone blocks from Eleusis that served as background for a series of figures carved from white marble.

NATURAL BUILDING MATERIALS: STONE AND MARBLE   9

Opposite. The Tabularium, a large building with a different order on each floor, was built during the first century B.C. on the eastern slope of the Capitoline Hill in Rome. In the foreground are the remaining white marble columns of the Temple of Vespasian and Titus.

high level of workability was evident in the rich decoration of the capitals of the Archaic-period Temple of Hera. In Sicily, Agrigento could extract stone to build its monuments from the hills surrounding the urban area. At Syracuse the eastern quarries and those of the Neapolis quarter remain impressive today for their majesty. It has been calculated that at the so-called Paradise Quarry alone, which includes the famous cave known as Dionysius's Ear, about 850,000 cubic meters of stone were extracted. Still visible in the quarries at Cusa, about 13 kilometers northwest of ancient Selinunte, are numerous calcarenite deposits originally destined to become column-drums in the gigantic peristyle of the city's so-called Temple G.

A particular use was made of dark limestone (Gr. *lithoi melainai*), whose colors ran from gray to blue. The best-known and most widespread of these was *lithos eleusiniakos*, which was extracted in the area of Eleusis in Attica, but other kinds of limestone were present in both Argolis (at Argos and Epidauros) and Phocis (at Delphi). Beginning in the fifth century B.C. these darker stones were used in architecture to create elegant dichromatic contrasts in conjunction with pale marble (as, for instance, in the frieze of the Erechtheum in Athens), or with the pale tones of other types of limestone (as in the toichobate of the Tholos of Epidaurus). At Olympia, for example, dark limestone was used in the Temple of Zeus to accentuate the floor of the central aisle and to emphasize the socle of the monumental cult statue created by Phidias.

Of inferior quality and thus of exclusively local use were the natural conglomerates (Gr. *arouraios*) and the breccias, all materials widely available in various regions of the Greek and Magna Graecia world. These sedimentary rocks result from the cementation of deposits of pebbles and gravel (puddingstone, or conglomerate rock), or of sharp-edged fragments that resulted from the shattering of earlier rock. The heterogeneous composition of these materials made them difficult to cut and work, and as a result they were usually employed for those parts of a construction that were not visible, such as foundations and filling behind walls. Beginning in the fourth century B.C., however, some of these stones were cut into square elements to be used, often together with limestone, in the visible walls of buildings (as at Delphi in the theater and *prytaneion*, the seat of government).

STONE IN ROME

In the second half of the first century B.C., the Roman architect and writer Vitruvius (*On Architecture* 2.7.1–2) discussed building stones and their uses, as well as the quarries from which both squared stone and supplies of rubble were extracted and supplied for buildings:

> Now these are found to be of unequal and unlike virtues. For some are soft, as they are in the neighborhood of the city at Grotta Rossa, Palla, Fidenae, and Alba; others are medium, as at Tivoli [Tibur], Amiternum, Soracte, and those which are of these kinds; some hard, like lava. There are also many other kinds, as red and black tufa in Campania; in Umbria and Picenum and in Venetia, white stone which indeed is cut, like wood, with a toothed saw. But all these quarries which are of soft stone have this advantage: when stones are taken from these quarries they are easily handled in working, and if they are in covered places, they sustain their burden.

Vitruvius notes, however, the dangers posed to soft stone by the weather and specific locations: "If they are in open and exposed places, they combine with ice and hoar frost, are turned to powder and are dissolved; along the sea coast, also, being weathered by the brine, they crumble and do not endure the heat."

As Vitruvius relates, the peoples of Rome and Latium had access to a wide range of stone materials for construction, varying in both quality and strength. For the most part these

NATURAL BUILDING MATERIALS: STONE AND MARBLE

The great Theater of Marcellus in Rome, begun by Caesar and completed by Augustus in 13 B.C., had a facade made entirely from blocks of travertine and was composed of three stories, each with a different order.

were tufas, pyroclastic coherent rocks that usually result from the cementation of deposits of volcanic lapilli. Tufas may have a very compact and strong structure, they may be porous, or, as in the case of pozzolana, they may be incoherent, that is loose or unconsolidated.

In Rome the first stone material to be used in architecture was that offered by the ground of the city itself: cappellaccio, a friable, granular, ash-gray tufa with a fairly fine grain. It remains to be established whether the stone was quarried from the area inside the Archaic settlement district of the early republic—perhaps on the slopes of the Capitoline at the Lautumiae, or quarry, at the site of the Tullianum—but quarries have been identified in the area of the Esquiline (near the Termini station, and near the church of Santa Bibiana), along the Via Tiburtina (at Vigna Querini), and along the Via Nomentana (at Villa Patrizi). Employed alongside red lithoid tufa, cappellaccio was the stone material principally used in Roman architecture of the sixth and fifth centuries B.C.; it was used to make the walls of the first stage of the Regia, the oldest stretches of the Servian Wall, and the most important temple in Rome, that dedicated to the triad of Jupiter, Juno, and Minerva on the Capitoline. The high friability of cappellaccio and its tendency to flake soon led the Romans to use stone materials that were more stable and exhibited superior aesthetic qualities. Even so, the use of cappellaccio was not abandoned; due to its high qualities of resistance to compression, it survived inside foundations until at least the second century B.C.

Following the conquest of Veii in 396 B.C., Rome began large-scale exploitation of the immense Grotta Oscura quarries, located about 4 kilometers outside Prima Porta. From these sites workers extracted a semilithoid porous tufa of grayish yellow, whose compactness made it possible to cut blocks about 59 centimeters high (equal to 2 Roman feet), twice the size of blocks possible from cappellaccio. Tufa from the Grotta Oscura quarries was used in Rome at many large work sites of the republican period, from the fortifications built in the fourth century B.C., after the attack of the Gauls, to Temples A and C of Largo Argentina, as well as two large bridges over the Tiber, the Aemilius and the Milvian. Extraction at the Grotta Oscura quarries declined near the end of the second century B.C., when other construction materials were introduced to Rome.

Another tufa widely used in the city's architecture was that of Fidenae, mentioned by Vitruvius and easily identifiable by its characteristic inclusions of black scoriae. The use of this stone, whose quarries were located near today's Castel Giubileo, probably began in 426 B.C., when the Romans conquered the city. Fidenae tufa appears in constructions on both the Palatine and the Capitoline, and it was used, together with tufa from Grotta Oscura, for the walls erected after the Gallic occupation of 390 B.C.

There were also pale-brown lithoid tufas, such as that from Monte Verde, quarried at the foot of the Janiculum and at Magliana, and that from the Aniene River, in various reddish tones, extracted from quarries at modern Tor Cervara. Both these materials came into use in Rome sometime in the third or second century B.C. Also available were the greenish or ash-gray peperino (Lat. *lapis tusculanus* or *albanus*) and the stone known as pietra sperone (Lat. *lapis gabinus*), similar but with a coarser grain. Peperino was mined in quarries at a relative distance from Rome (about 25 kilometers), in the area of Marino, Castelgandolfo, Albano Laziale, and Ariccia. Pietra sperone came from the area of ancient Gabii, about 15 kilometers from Rome; transfer of this stone was facilitated by the presence of a water route that made possible river transport.

Between the end of the third century and the beginning of the second century B.C., another kind of stone made its entrance in Roman architecture, and was destined to assume a dominant position among construction materials. This was travertine (Lat. *lapis tiburtinus*), a sedimentary, white limestone that is still frequently quarried in the zone of Bagni di Tivoli. Vitruvius (2.7.2) relates several qualities of this stone: "Travertine, however, and all stones which are of the same kind, withstand injury from heavy loads and from storms; but from fire they cannot be safe; as soon as they are touched by it, they crack and break up." Despite its tendency to calcinate in contact with fire, travertine offered the Romans numerous advantages: attractive aesthetic qualities, ease of extraction and working, high mechanical capacity, and excellent resistance to atmospheric agents. These characteristics propelled architects of the Late-Republican and imperial periods to widespread use of the stone, preferring travertine when marble was not available. The importance of travertine to Roman architecture can be seen in its use for major buildings designed for spectacle, including the Theater of Marcellus, the Flavian Amphitheater, and the Stadium of Domitian.

Of superior quality, but difficult to work because of its characteristic hardness, is leucite, a fine-grained, porphyritic compact rock of igneous origin that was probably mined at Capo di Bove, along the Via Appia. Identified by some with the *lapis silex* mentioned by ancient sources, leucite was used particularly in the form of cobbles for paving roads (as along the Via Sacra), and in the production of dark-colored mosaic tesserae. Today it is still employed to make the cobblestones used for the historic streets of Rome.

THE BUILDING STONES OF ROMAN ITALY

As in Rome, other centers on the Italian peninsula made use of locally available stone materials. This often led to the development of specialized skills in working each type found at an individual locality, especially when a stone was used for sculpting the decorative elements, such as capitals and trabeations.

Just as Vitruvius lists the kinds of stone that were typical of Rome's individual *regiones*, such as the dark tufa of Campania or the pale limestone of Venetia, he also recounts (2.7.3–4) details of quarries near Tarquinii, a source for *lapidicinae Anicianae*, which has a color similar to that of Alba. These quarries were located primarily around the lake of Bolsena, and also in the prefecture of Statonia, and the stones extracted there, claimed Vitruvius, had "infinite virtues; for they can neither be injured by weathering under frost nor by the approach of fire. But the stone is firm and wears well over a long time.... This we may especially judge from the monuments, which are about the municipality of Ferentum, made from these quarries. For they have large statues strikingly

View of the remains of the amphitheater of Lupiae (modern Lecce), constructed using large blocks of local argillaceous limestone.

made, and lesser figures and flowers and acanthus finely carved. These, old as they are, appear as fresh as if they were just made."

At Pompeii a type of stone known as cruma was mined from the same layer of volcanic origin on which the city itself came to stand. This was a light, porous material, whose poor strength was balanced by its ease of cutting and working. For this reason cruma was widely used as a construction material at the work sites of the Samnite period and the late republic, but it was eventually replaced by stones of better quality. As early as the sixth century B.C. another kind of stone had been introduced to Pompeii, known as Sarno limestone and quarried along the banks of the river of that name. A stone with a porous nature, it belongs more properly to the class of the travertinic conglomerates than to limestone.

Relatively harder and stronger was the dark tufa today known as pappamonte, used in construction of the city's oldest fortifications. Pappamonte had the advantage of increasing its resistance to atmospheric agents following exposure to air when quarried. It was used to make not only numerous walls in the city but also many decorative elements, such as some of the Doric capitals from the Archaic period in the temple of the Triangular Forum. However, Pompeii stonecutters soon turned to the gray tufa of Nocera, as well as to the paler and yellowish tufa from the Phlegraean Fields, for the creation of more refined decoration. Nocera tufa, easily recognizable by the presence of small black scoriae within the rock, was the material most often used in Pompeian architecture of the second century B.C., when the city became a rich center of Hellenistic culture.

A section of richly decorated cornice from excavations at Aquileia. The monuments of this city made large-scale use of Aurisina, a limestone from quarries in ancient Nabresina, near Trieste.

In the region of Apulia et Calabria, the city of Lupiae (modern Lecce) had abundant quarries of an excellent pale-colored argillaceous limestone that was easily worked and also hardened upon exposure to air. At Lupiae it was used, for example, to build the amphitheater in large, squared blocks, as well as for smaller tesserae within the curtain walls of *opus reticulatum*.

In Etruria the Romans, as before them the Etruscans, were able to exploit large tufa deposits at Bolsena, Bracciano, and Vico; the limestone quarries at Grossetano and in the areas of Arezzo and Siena; and limestone and marly calcareous stone available in various parts of the territory.

The region of Venetia et Histria also possessed various types of stone. A compact, fine-grained limestone of a cream-yellow color was quarried in the area around Brixia (modern Brescia); known today as Botticino, it remains in use. Together with imported marbles, this was the only stone used in construction of the city's theater, including the squared stones of *opus quadratum* and architectural decoration, erected at the end of the first century B.C. Equally widespread was the use of various types of nodular limestone found in the area of Verona, from the type known as Biancone to those stones characterized by a brighter chromatic range, red to pink. In addition to the use of Biancone in constructions at Verona itself (for the theater, amphitheater, Porta dei Borsari, and Gavi Arch), it also was exported throughout the Po area. At Aquileia buildings used the "stone of Aurisina," a gray limestone, strong and easy to cut, that was mined in the quarries at ancient Nabresina, near Trieste.

Opposite. The Propylaea was designed by Mnesicles and erected between 437 and 432 B.C. to monumentalize the western gateway to the Athenian Acropolis. It was built entirely in Pentelic marble with the exception of some terracing structures, which used limestone blocks crowned by a euthynteria in gray-blue stone from Eleusis.

In northwest Italy, in the area of Transpadana at the ancient Augusta Praetoria (modern Aosta), several deposits of travertines and puddingstone conglomerates were available in the valley, although the city built its amphitheater using blocks of a gray limestone from Aymavilles, a town south of the city. For the construction of Turin's city walls and theater, however, the choice was made to use a gneissic stone from the Val di Susa, a material that was used, as one would expect, in Susa itself (ancient Segusium).

THE WHITE MARBLES OF GREECE AND ASIA MINOR

Unlike the materials discussed thus far, marbles are rocks with a crystalline structure generated by the metamorphosis of carbonatic stones such as limestone and dolomitic limestone. In the ancient world white marbles were the material preferred not only by great sculptors but also by leading architects, first Greek and then Roman.

Used from the end of the Geometric period in places where it was the material of ready availability, marble began to be incorporated sporadically in Greek architecture during the sixth century B.C., and appeared more widely in the following century. Describing the region of Attica, the Greek scholar Strabo (*Geography* 9.1.23) recalled that "of the mountains, those which are most famous are Hymettus, Brilessus, and Lycabettus; and also Parnes and Corydallus. Near the city are most excellent quarries of marble, the Hymettian and Pentelic." Athens thus had two different types of easily accessible marble. The first was the Hymettian (Gr. *hymettios lithos*, Lat. *marmor hymettium*), a stone with a finely grained structure and bluish white color, often with darker stripes. It was quarried on Mount Hymettus, about 11 kilometers southeast of the city. The second was Pentelic (Gr. *pentelesios lithos*, Lat. *marmor pentelicum*), a pale, finely grained marble with warm tones, often with subtle veins of a bright greenish color. It came from quarries located on Mount Pentelicus, about 14 kilometers northeast of Athens. These two types of marble constituted the primary construction materials for the great program undertaken by Pericles in the second half of the fifth century B.C., but they had been used only occasionally in Athenian temples of the previous century. The white marbles quarried on two of the major islands of the Cycladic archipelago, Paros and Naxos, had been used in statuary since the seventh century B.C., and were incorporated in several major constructions of the Archaic period.

Parian marble (Gr. *parios lithos*, Lat. *marmor parium*), one of the most celebrated stones of antiquity, has a finely grained structure and a snow-white color tending to gray and pale yellow. The island furnished various qualities of marble, the most prized of which was extracted from quarries in mines near Stephani on Mount Marpissa, northeast of Paros's capital city. Ancient sources refer to the marble as *lychnites lithos* because, as Pliny the Elder states (*Natural History* 36.14), "it was quarried in galleries by the light of oil lamps"—*lychnites* meaning "lamplike." Another explanation is that the term referred to the marble's distinctive translucence, which in sunlight seems to radiate a glow from inside the stone. Other quarries, such as that at Lakkoi, were worked in the open air to extract blocks, which were occasionally very large. Some of these pieces today lie where they were quarried, abandoned. Parian marble was used early in architecture not only on the island but also at many work sites in continental Greece. As early as the sixth century B.C., it was imported to Athens to be used in the creation of the metopes and elements of the cyma for the original Parthenon on the Acropolis. Of particular documentary value is the information provided by Herodotus (*Histories* 5.62–63) regarding reconstruction of the Temple of Apollo at Delphi at the end of the sixth century B.C. According to Herodotus, for the building's main facade and its pedimental statues, the family of the Alcmaeonidae—in charge of construction—chose to replace the

NATURAL BUILDING MATERIALS: STONE AND MARBLE 17

Opposite. View of the entrance to the Marathi marble quarry on the island of Paros.

limestone called for in the original design with white Parian marble. The translucency of all Parian marble extended up to a thickness of 3.5 centimeters, and its use for the roof tiles of certain temple buildings made possible the passage of light to the interior, giving the structures a particularly evocative atmosphere.

The nearby island of Naxos was rich in an equally excellent white marble, often recognizable by the larger size of its crystals. It was quarried in several places on the island, including Apiranthos, Kinidaros, and Komiaki. Even today, to the north of Komiaki in the area known as Apollonas, there is a colossal statue of a bearded Dionysius (10.5 meters high), left unfinished by the stonecutters and abandoned at the quarry. Another male statue (Gr. *kouros*), 6.4 meters high, had a similar destiny, and since the sixth century B.C. it has been lying in the valley of Melanes near Fleio. As early as the Archaic period the Naxians were already exporting their island's marble, and the technical skills they rapidly acquired are evident in various works they created—the Oikos and the Kouros of the Naxians on Delos, as well as the Column of the Naxians at Delphi. Thanks to the presence of deposits of corundum, an abrasive sand quarried in the area of Apiranthos and Koronos, the Naxians soon came to specialize in cutting large blocks of marble. Pliny the Elder (*N.H.* 36.51) describes the technique used to cut marble and thus "carve up luxury into many portions": "The cutting of the marble is effected apparently by iron, but actually by sand, for the saw merely presses the sand upon a very thinly traced line, and then the passage of the instrument, owing to the rapid movement to and fro, is in itself enough to cut the stone." Writing during the Roman imperial period, Pausanias (*Descr. of Greece* 5.10.3) attributed the invention of marble tiles to a Naxian artisan by the name of Byzes.

Together with these white marbles, others came into use during the Archaic period for the architecture and sculpture of Greece in Asia Minor: Proconnessian marble (Gr. *prokonnesios lithos*, Lat. *marmor proconnesium*), quarried in various areas on the island of Marmara; the marbles of Ephesus and Miletus; those quarried in the area of Herakleia at Latmos; and those from the quarries at Usak. During this period continental Greece and its islands began to use white Doliana marble, quarried on Mount Parnon in Arcadia, as well as that from the quarries at Tenaro in Laconia, from Kerki on Samos, and naturally from Thasos (Gr. *thasios lithos*, Lat. *marmor thasium*), which Pliny the Elder believed rivaled the marbles of the Cyclades. Exploitation of the Docimian quarries in Phrygia in Asia Minor began in the fifth century B.C., but the Greeks did not begin to quarry the marble of Aphrodisias in nearby Caria until the Hellenistic period.

THE FIRST GREEK ARCHITECTURAL USES OF WHITE MARBLE FROM THE GREEK ISLANDS

During the last few decades archaeological discoveries on Naxos, the largest of the Cycladic islands, have added new information about Greek architecture of the Archaic period, revealing the most ancient methods applied to the use of marble in temple constructions.

At Yria, in the area where myths located the birth of Dionysius, archaeologists working outside the city have uncovered the remains of a large sanctuary, dated to the beginning of the eighth century B.C. and furnished with a small, rectangular oikos temple (about 5 by 10 meters) that faces south. The building was constructed with a low socle of stone chips and an upper wall of unfired bricks. A row of three wooden columns divided the interior of the temple into two aisles, and at the center supported the weight of a roof that was probably flat.

In the second half of the century, perhaps around 730 B.C., the temple was rebuilt at a larger size (about 11 by 16.5 meters). The original planimetric scheme underwent notable changes: the interior was articulated into four aisles by three rows of wooden columns, and continuous benches ran along the internal

NATURAL BUILDING MATERIALS: STONE AND MARBLE

Excavations have uncovered the remains of the great Sanctuary of Dionysius near the modern village of Yria on the island of Naxos. Shown here are elements from the last period of the temple, dated to the middle of the sixth century B.C.

walls of the hall for the seating of participants in the ritual meal that took place around the hearth (Gr. *eschara*). The techniques used to build the temple marked an important stage in the evolution of construction, as they were distinguished by the use of local granite for the walls. Alongside this material the island's white marble makes its first appearance, most significantly in the bases of columns and the cornerstones of walls, giving the structure a distinctive, dichromatic appearance.

The temple underwent an extensive period of monumentalizing during the first half of the seventh century B.C. The position of the building's side walls did not change, but the axial colonnade was eliminated, and the interior space was divided into three aisles. With the addition of an anterior prostyle portico of the tetrastyle type, the temple reached a length of 19.7 meters. The columns continued to be constructed with wooden shafts, but now rested on white marble bases and probably had marble capitals. The roof of the building represented a distinctive technical solution, as it was shaped so that rainwater would flow into gutters that were also made of Naxian marble.

A final important architectural touch took place around the middle of the sixth century B.C. The temple, its cella now divided into three aisles, reached a notable size (13.5 by 28.3 meters) with the addition of a monumental pronaos and an adyton. During this phase the builders used pieces of local granite for the perimeter walls, but chose local white marble to make all the other parts of the building, from the Ionic columns to the roof, which was given a covering of tiles of the insular Corinthian type, decorated on the sides with palmette antefixes.

Aside from the temple of Yria, the building that best illustrates the mastery in the use of marble achieved by the island builders is the Oikos of the Naxians, erected by the citizens of Naxos at the important Sanctuary of Apollo on Delos. Positioned near the southern entrance to the sacred zone, the monument

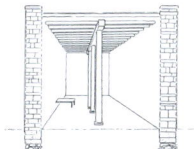
Stage I (early 8th century B.C.)

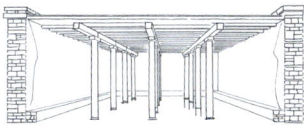
Stage II (ca. 730 B.C.)

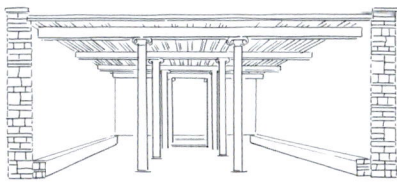
Stage III (first half of 7th century B.C.)

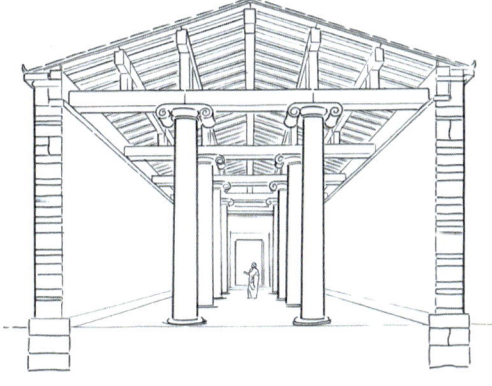
Stage IV (mid-6th century B.C.)

Left. The complex layers of the temple at the Sanctuary of Dionysius at Yria reveal the different stages of construction, from the first building at the beginning of the eighth century B.C. to the great expansion at the middle of the sixth century B.C.

Below. Detail of the roof of the temple of the Sanctuary of Dionysius at Yria, composed of elements made using the island's white marble.

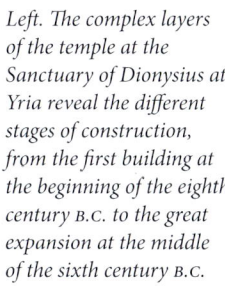

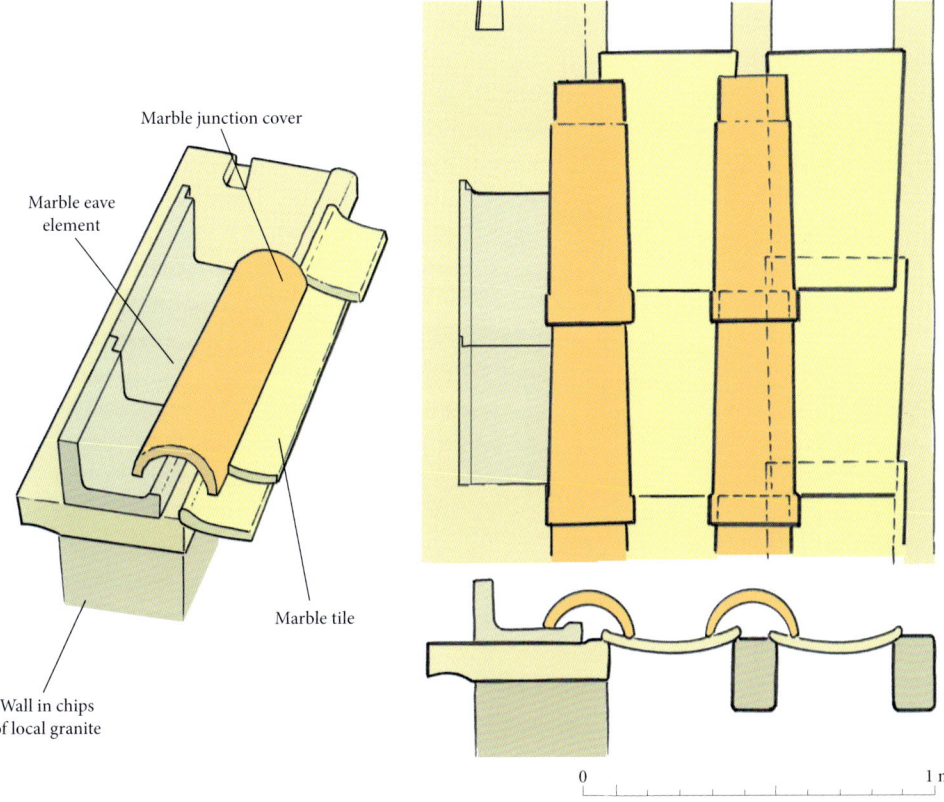

Marble junction cover
Marble eave element
Marble tile
Wall in chips of local granite

Above. View from the sea of the buildings on the island of Delos.

was probably built in the early decades of the sixth century B.C. as a space for assembly and ritual meals (Gr. *hestiatorion*). The rectangular building (roughly 10 by 24 meters) faces west, with its interior divided in two aisles by a series of eight columns; on the front facade was a tristyle portico *in antis*. During a second phase, dated around 550 B.C., an Ionic prostyle tetrastyle portico was added to the east facade.

The Oikos displays the expertise of Naxian builders, who by then specialized in the use, mining, and transport of marble. Their abilities are also demonstrated by the monumental male statue immediately north of the building: a monolithic sculpture of Naxian marble about 9 meters high, set on a monolithic base, also of Naxian marble, more than 5 meters long and weighing approximately 32 tons. After transporting these pieces to Delos, the inhabitants of Naxos dedicated them to Apollo and proudly inscribed on the base, "I am of the same stone, both statue and base."

The building had a foundation in simply shaped blocks of local granite, while the walls were made with large pieces of the same material and elongated pieces of gneiss (a stone similar to granite but schistose). The Naxian

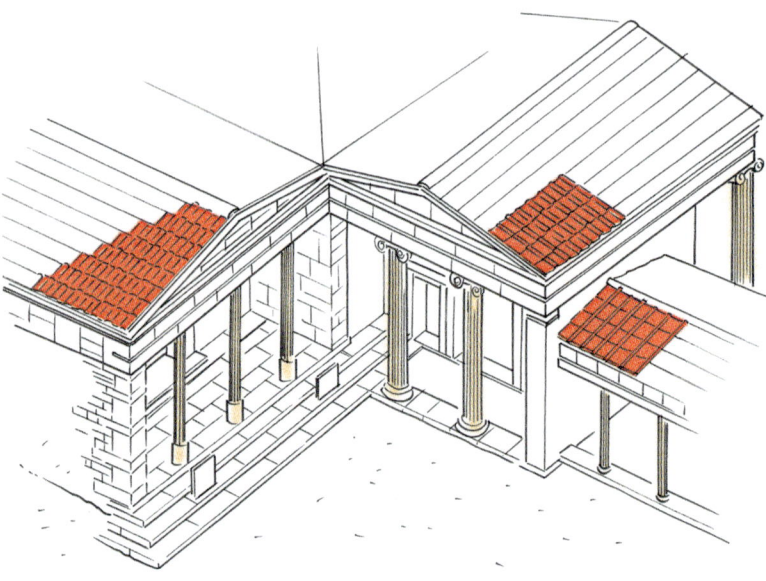

builders reinforced the walls at their corners by inserting larger blocks, creating the impression of stone pillars. What remains of the northern door is a threshold made from a block of white marble. The same material was used for the rest of the building's elements, beginning with the internal supports. The bases of the columns were composed of a sort of lower disk (roughly a meter in diameter) and an upwardly tapering cylinder; the shafts were articulated with twenty-four sharp-edged grooves, while the capitals were of the large-volute Ionic style.

The most notable aspect of the Oikos was undoubtedly the structure's roofing system. From the elements of the framework to those of the roof itself, everything was made in Naxian marble. A course of marble blocks crowned the top of the walls, forming a sort of smooth frieze; on top of the frieze rested a row of sculpted slabs that bore openings into which the roof beams were inserted. The roof was made of slightly concave tiles and convex junction covers, all in pale marble. Along the line of the eaves the covers terminated in semicircular marble antefixae decorated with incised Gorgon faces.

The use of white marble more clearly marked the second period of the monument's construction, when the tetrastyle *in antis* prostoon (or porch) was added to the western side. This new structure—including a three-step crepidoma, columns, the antae of the walls, trabeation, and roof—was built entirely in Naxian marble.

There is no doubt that the creation of such a complex building system required a well-organized work site, both in terms of the initial formal definition of individual elements and in the direction of the work necessary to assemble the building. Traces of scaffolding used to raise the columns and to position the roof elements have been found inside the structure, revealing that boards were set on top of wooden poles that had been fixed in the ground, and the boards were then inserted in putlog holes along the sides of the columns.

Above. Early in the sixth century B.C., at the Sanctuary of Apollo on Delos near the southern entrance to the sacred precinct, the inhabitants of Naxos built the large building known today as the Oikos of the Naxians. Beside it they erected a monumental male statue (kouros) about 9 meters high, set on a monolithic base more than 5 meters long.

Opposite, bottom. Drawing of the appearance of the southern entrance area of the temenos of the Sanctuary of Apollo around the middle of the sixth century B.C., with the large propylon inserted between the Oikos (left) and the Stoa (right) of the Naxians.

NATURAL BUILDING MATERIALS: STONE AND MARBLE

This Luna-marble frieze, decorated with dolphins and tridents, embellished the great hall of the Basilica of Neptune, which was reconstructed in the period of Hadrian and located against the rear wall of the Pantheon in Rome.

MARMOR LUNENSE

They say that the maximum length of Tyrrhenia—the coastline from Luna as far as Ostia—is about 2,500 stadia. . . . Luna . . . is a city and also a harbor, and the Greeks call the city as well as the harbor "Harbor of Silene." The city, indeed, is not large, but the harbor is both very large and very beautiful, since it includes within itself several harbors, all of them deep up to the very shore. . . . The harbor is shut in all around by high mountains, from which the high seas are to be seen, as also Sardo [Sardinia], and a considerable stretch of the shore on either side. And the quarries of marble, both white and mottled bluish-gray marble, are so numerous, and of such quality (for they yield monolithic slabs and columns), that the material for most of the superior works of art in Rome and the rest of the cities is supplied therefrom; and indeed the marble is easy to export, since the quarries lie above the sea and near it, and since the Tiber in its turn takes up the cargo from the sea and conveys it to Rome.

So wrote Strabo (*Geography* 5.2.5) during the Augustan period, in his description of the city of Luna (modern Luni), referring to one of the most important decorative materials of Roman architecture.

Luna marble (Lat. *marmor lunense*) is still quarried today in the chief marble basins at Colonnata, Miseglia, and Torano. A white marble with a fine compact grain and of excellent workability, it is luminescent but much less translucent than marbles from the east. It may present shades of cerulean blue or, as in the case of the Calacatta variety, golden white tones with a slight veining ranging from golden yellow to pale green.

Employed as early as the mid-first century B.C. in the *domus* ("house") of Mamurra, prefect of Caesar, Luna marble came into large-scale use in Rome in 37 B.C., as part of the material employed for reconstruction of the Regia. Its rapid diffusion in the Augustan period was immortalized by the historian Suetonius (*Augustus* 28), who referred to the building programs carried out by Rome's first emperor: "Since the city was not adorned as the dignity of the empire demanded, and was exposed to flood and fire, he so beautified it that he could justly boast that he had found it built of brick and left it in marble." Beginning under the emperor Tiberius, the quarries—by then state property—became the center of a complex system of extracting and distributing marble, controlled by the imperial family and administered by members of the empire's accounting office (Lat. *tabularii marmorum lunensium*).

The close vicinity of the sea facilitated transportation of Luna marble, and when blocks, already half-dressed at the quarry, arrived at the Portus Lunae, they were loaded on large ships (Lat. *naves lapidariae*). The primary route was undoubtedly toward the port of Ostia, where there was an immense storehouse (Lat. *statio marmorum*). From there the material traveled to Rome by way of the Tiber, and was then taken to a collection and storage center established along the riverbanks at the foot of the Aventine. Shipments identified in numerous wrecks discovered in the Mediterranean have offered important information concerning the circulation of Luna marble in the western provinces. A ship that sank off Saint-Tropez carried architectural elements for a large temple with columns more than 13 meters high, similar in size to the column drums that constituted the cargo of a wreck from the Tiberian period found at Porto Novo in southern Corsica.

IMPORTED COLORED STONES AND MARBLES

By the first century A.D. the import of white marbles—in particular those from Paros, Proconessus, and Thasos—to Rome from the Orient was decreasing. The progressive introduction of Luna marble was accompanied (and in many cases preceded) by the appearance of other stone materials, which because of their chromatic range immediately acquired a decorative function. Already in the Augustan period, Strabo (*Geography* 9.5.16) wrote about marble quarries on the Greek island of Skyros, and attributed the decline in value of white marbles in Rome to the import of monolithic columns and large slabs of colored marble. During the imperial period many colored stones and marbles, originating at a great number of quarries in the Mediterranean, were used in Roman architecture. The Edict on Maximum Prices issued by the emperor Diocletian in A.D. 301 lists nineteen different types of marble, of which only three or four were white.

The principal types of stone are listed below by area of provenance, together with individual quarrying sites, their characteristics, and their principal uses in architecture. These stones include not only sedimentary and metamorphic rocks, such as those discussed earlier, but also magmatic, that is, naturally intrusive (plutonic) or extrusive.

Colored Stones and Marbles from Greece

Stones of Sedimentary Origin
• Chios breccia (Lat. *marmor chium*), also called portasanta, mined on the island of Chios. This is a breccia with a variable appearance, often red with clasts of paler tones.

Opposite. The interior of the Pantheon, reconstructed during the reign of Hadrian to replace a temple erected by Augustus, still testifies to the richness of ancient buildings. Despite multiple transformations and restorations, it preserves much of its original decoration, made using marble from quarries located throughout the Mediterranean.

Stones of Magmatic Origin

- Serpentine (Lat. *marmor lacedaemonium*), also called porphyry verde antico, mined near the ancient village of Krokeai, not far from Sparta, in the Peloponnesus. Said by Pliny the Elder (*N.H.* 36.55) to be "brighter than any other," it is characterized by a dark green color with lighter phenocrysts. Together with red Egyptian porphyry, it was the most costly construction stone in the Roman world.

Stones of Metamorphic Origin

- Settebasi (Lat. *marmor scyreticum*), from quarries on Skyros. This stone presents a variety of colors, from brown to whitish with clasts that range from very pale tones to red. The most prized variety is called Semesanto.
- Cipollino verde (Gr. *karystios lithos*, Lat. *marmor carystium*), mined near ancient Karystos. This marble appears in various shades of green with darker undulating or parallel veins. Like cipollino rosso, it was named for its resemblance to an onion (*cipollino*) when sliced crosswise. It was widely used for monolithic column shafts and for slabs of wall dressing.
- Fior di pesco or persico (Lat. *marmor chalcidicum*), mined not far from Eretria (near ancient Chalcis) in Euboea. It is pink, sometimes grayish, generally with veins and spots. Introduced to Rome in the Augustan period, this stone was used to make various types of architectural elements.
- Rosso antico (Gr. *tainarios lithos*, Lat. *marmor taenarium*), mined at various localities on the Mani peninsula, in the Peloponnesus. As indicated by its name, it was a red marble, sometimes with thin dark veining. A favorite in statuary with Dionysian subjects because of its wine-colored tones, rosso antico was used for cornices, shafts, and capitals of columns and as dressing slabs.
- Verde antico (Gr. *atrakios lithos*, Lat. *marmor thessalicum*), mined on Mount Mopsius in Thessaly. This marble has a characteristic green color, with marks of darker color or black and others of white.

Colored Stones and Marbles from Asia Minor

Stones of Sedimentary Origin

- Alabastro fiorito (Lat. *marmor hierapolitanum*), in several known varieties, one of which was mined near Hierapolis in Phrygia, in Turkey. It has a light color with generally darker spots similar to arborescences.
- Breccia corallina (Lat. *marmor sagarium*), from quarries at Vezirhan (in ancient Bithynia) in Turkey. This is a pink limestone breccia, with white or pink clasts. Probably introduced to Rome in the Augustan period, it was used for column shafts or, as in the case of several homes at Pompeii and Herculaneum, in the form of dressing slabs.
- Africano marble (Lat. *marmor lucullaeum*), mined, despite its modern name, at Sigacik, near ancient Teos, in Turkey. This breccia is formed of light-colored clasts shading from pink to red and immersed in a dark cement. It takes its Latin name from the consul Lucius Licinius Lucullus, who introduced this marble to Rome during the first century B.C.
- Occhio di pavone, peacock's eye (Lat. *marmor triponticum*), quarried near the modern village of Kutluca, not far from ancient Nicomedia, in Turkey. The modern name is derived from the stone's appearance, reddish pink with white macrofossils that, when cut crosswise, have a circular shape similar to the "eyes" in the tail of a peacock.

Rocks of Magmatic Origin

- Granito violetto, violet granite (Lat. *marmor troadense*), mined on the slopes of the Cigri Dag in the Troade, located a short distance from ancient Neandria. It presents a gray color (with darker and lighter varieties) together with medium-size grayish violet spots. Granito violetto arrived in Rome only at the beginning of the second century A.D. but it soon enjoyed great popularity throughout the empire, and was used especially for monolithic columns and large pillars.

Opposite. The columns of the pronaos of the Temple of Antoninus and Faustina, in the Roman Forum, were made with monolithic shafts in cipollino verde marble from Euboea.

Rocks of Metamorphic Origin

- Cipollino red (Lat. *marmor iassense* or *marmor carium*), mined near ancient Iasos in Turkey. It appears in a red color with lighter veining, in a type with a uniform color and, more rarely, as a breccia. Used locally in the Hellenistic period, cipollino red had widespread use in the imperial period, especially in the central-eastern Mediterranean.
- Pavonazzetto (Gr. *phrygios lithos*, Lat. *marmor phrygium* or *marmor docimenium*), mined at quarries in Docimium (modern Ischehisar) in Phrygia. This is a fine-grained white marble with violet spots. In Diocletian's edict it is cited as the most costly marble after red porphyry and serpentine. Because of its aesthetic qualities the architectural use of pavonazzetto spread rapidly throughout the empire.

Colored Stones and Marbles from Africa

Stones of Sedimentary Origin

- Alabastro a pecorella, so called for its appearance similar to the fleece of a sheep. It was mined in quarries at Ain Tekbalet in Algeria, and introduced to Rome somewhat late, between the first and second centuries A.D.
- Egyptian alabaster (Lat. *lapis alabastrites*), also called cotognino alabaster, mined along the east bank of the Nile. Honey colored, it has stripes and spots of white, pink, and brown.
- Red and yellow breccia (possibly Lat. *lapis knekites*), also called breccia corallina ombrata, mined along the Nile Valley in the Wadi Abu Gelbana. This is a red breccia with clasts of pale yellow or gray; in Roman construction it was used primarily as floor slabs.
- Giallo antico (Lat. *marmor numidicum*), mined in quarries at Simitthus (modern Chimtou) in Tunisia. Yellow in color, with light and dark tones, it is found in veined and brecciated forms. Giallo antico was among the first stones imported to Rome, around 100 B.C., and despite its high cost was widely used to make columns and other architectural members, floor slabs, and wall dressings.
- Numidian nero antico (Lat. *lapis niger*), mined at quarries in Simitthus and Gebel Aziz. This stone is an attractive black with whitish veins and spots. A similar limestone was mined in Greece (at Chios and Cape Tenaro) and Bithynia (at Adapazari), and at the quarries in Palombino in Latium.

Rocks of Magmatic Origin

- Egyptian diorite (Lat. *lapis thebaicus*), also called black Egyptian granite and characterized by a black or dark gray color with whitish pink spots. Mined at Gebel Nagug, to the south of Syene (modern Aswan), it was used to make monolithic columns, large pillars, and slabs for dressing.
- Porphyry rosso antico (Gr. *porphyrites lithos*, Lat. *lapis porphyrites*), mined at Mons Porphyrites (modern Gebel Dokhan) in the eastern desert of Egypt. This was one of the two most costly marbles sold during the Roman period. Purple in color, with white-pink or black phenocrysts, it not only was chosen for architectural columns and pillars but also was a preferred material for imperial statues and sarcophagi (of Helena and of Saint Constance, for example). There were two other colors of Egyptian porphyry, green (Lat. *lapis hieracites*) and serpentine (black), both from quarries at Gebel Dokhan.
- Syenite (Gr. *syenites lithos*, Lat. *lapis pyrrhopoecilus*), also called red or pink Aswan granite, mined along the banks of the Nile near ancient Syene. Found in colors that run from pink to dark red, it was used in Roman architecture to make large monolithic columns, pilaster strips, and dressing slabs. The monumental shafts of the internal columns of the pronaos of the Pantheon in Rome are made of syenite.

Rocks of Metamorphic Origin

- Green Egyptian breccia (Gr. *hekatontalithos lithos*), mined in the Wadi Hammamat near ancient Mons Basanites. This is characterized by a puddingstone conglomerate composed of rounded parts of diverse origin that are joined by green cement.

NATURAL BUILDING MATERIALS: STONE AND MARBLE 29

Prices of marbles and colored stones on the list issued by Diocletian in A.D. 301

Name on edict	Modern name	Price per cubic ft (29.6 cm per side)
lapis porphyrites	porphyry rosso antico	250 denari
marmor lacedaemonium	serpentine, porphyry verde antico	250 denari
marmor numidicum	giallo antico	200 denari
marmor lucullaeum	africano marble	150 denari
lapis pyrrhopoecilus	syenite, Aswan red granite	100 denari
marmor claudianum	granito del foro	100 denari
lapis alabastrites	alabastro egiziano	75 denari
marmor docimium	pavonazzetto	200 denari
euthydemianos lithos	unidentified	60 denari
anakastenos lithos	unidentified	40 denari
marmor triponticum	peacock's eye?	75 denari
marmor thessalicum	verde antico	150 denari
marmor carystium	cipollino verde	100 denari
marmor scyreticum	Skyros white marble	40 denari
herakleiotikos lithos	marble from Herakleia at Latmos?	75 denari
marmor lesbium	Lesbos white marble	40 denari
marmor thasium	Thasos white marble, Greek black marble	50 denari
marmor proconnesium	marmo cipolla	40 denari
marmor potamogallenum	unidentified	40 denari

- Granito del foro (Lat. *marmor claudianum*), mined at Gebel Fatireh (ancient Mons Claudianus) in the eastern Egyptian desert. It presents black crystals on a white ground, and was used for large columns and dressing slabs. Many shafts in granito del foro made for the architect Apollodorus of Damascus to use in the aisles of the Ulpian Basilica, in Trajan's Forum, still exist today.
- Greco scritto, probably mined in Algeria near Annaba. This is a white marble with blue-gray spots and short veining, introduced to Rome around the end of the first century A.D.

QUARRIES AND METHODS OF MINING STONE AND MARBLE

The creation of a building in antiquity generally entailed a variety of stone materials, as architects took advantage of the static capabilities of individual types of stones to serve different purposes, and as the availability of individual stones varied. Each kind of stone

One of the rock faces of the great white-marble quarries on the island of Thasos. Visible are fault lines within the marble, used to facilitate the extraction of blocks, as well as small pieces that remained after working the blocks.

was priced differently, depending on its aesthetic qualities, the difficulty of working it, and most of all on the distance between the quarry and the site of use.

The working and mining of a given stone constituted a single, closely related concept, and artisans became so specialized at working a particular type that they were called to construction sites upon the arrival of transported blocks. For example, several inscriptions concerning the reconstruction of the Temple of Apollo at Delphi during the fourth century B.C. refer to the presence of Peloponnesian stonecutters who had been summoned for their experience with limestone. Similarly, artisans skilled at working the pale marble of Paros accompanied blocks when they were shipped from the quarry, even to sites as distant as Magna Graecia, where they sculpted the tiles used in several large temples, including those at Caulonia, Croton, and Metapontum. The same specialized approach applied to stonecutters working during the Roman period. Between the end of the second and beginning of the third centuries A.D., the emperor Septimius Severus began a large program for the architectural renewal of the city of his birth, Leptis Magna in Libya, and he summoned skilled workers from Attica and Asia Minor together with the white marbles specially imported from Attica and the Peloponnesus. One of these marble artisans, a certain Eleuseinios from Attica, left his signature on the base of a Pentelic marble column.

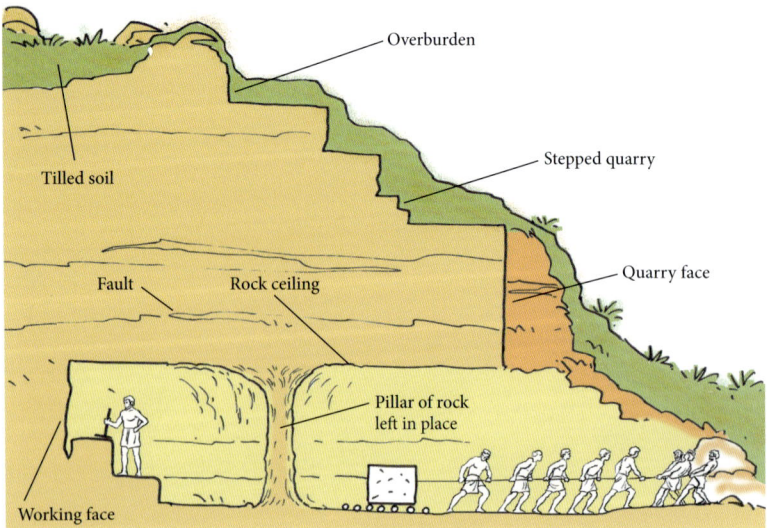

Top. Stone or marble was extracted by way of an open-pit, or a stepped, quarry, or by digging a mine.

Bottom. Cultivation of a quarry gallery with the use of metal wedges for detachment of blocks.

Every structure built with stone has its beginning in a quarry (Gr. *latomia* or *lithotomia*, Lat. pl. *lapidicinae*), often located far from the construction site. All the material destined to be worked and transformed into the various stone elements of a building came from a quarry. Pliny the Elder (*N.H.* 36.1–3) reflected on the use of stones from distant places in architecture and sculpture:

> Mountains, however, were made by Nature for herself to serve as a kind of framework for holding firmly together the inner parts of the earth, and at the same time to enable her to subdue the violence of rivers, to break the force of heavy seas.... We quarry these mountains and haul them away for a mere whim; and yet there was a time when it seemed remarkable even to have succeeded in crossing them.... We remove the barriers created to serve as the boundaries of nations, and ships are built specially for marble. And so, over the waves of the sea, Nature's wildest element, mountain ranges are transported to and fro.... When we hear of the prices paid ... when we see the masses of marble that are being conveyed or hauled, we should each of us reflect, and at the same time think how much more happily many people live without them.

The exploitation of a quarry began with identification of the quality of the material available to be mined. Some quarries were of antique origin; others were the fruit of an accidental discovery. Vitruvius (10.2.15) relates a curious anecdote concerning the fortunate discovery of the quarries at Ephesus by Pixodarus, a shepherd who lived in the area:

> Now when the citizens of Ephesus planned to build a temple [the Artemision] of marble and decided to obtain marble from Paros, Proconnesus, Heraclea, and Thasos, Pixodarus was driving his sheep and was pasturing them in the same place. And there two rams, butting together, overran one another, and, in the rush, one of them struck a rock with his horns and a chip of the whitest color was thrown down. So Pixodarus is said to have left his sheep on the hills and to have run with the chip of marble to Ephesus at the time when there was a great discussion about the matter. Thus the citizens decreed him divine honors and changed his name: instead of Pixodarus he was to be named Evangelus ["Good Messenger"].

After identification of the area from which good-quality material could be obtained, a quarry could be exploited using one of two systems. The most widespread method employed was that of an open-pit, or surface, quarry, which had already been adopted in Egypt during the Old Kingdom (early third millennium B.C.). Stone was mined at the surface following a preparatory stage (Gr. *anakathairō*), during which the outermost layer, known as the overburden, was eliminated. This layer is composed of the deposit of earth and its related plants along with the outermost section of bedrock; stone from the latter is of poor quality and thus unusable except as rubble and rocks for fillings. The development of the quarry advanced from the top down, beginning at the summit of the rock vein, known as the head, and continued with the progressive removal of roughly parallelepiped elements—that is, each face was a parallelogram—resulting in the formation of a stepped quarry (Gr. *katatome*).

Left. Blocks of white marble in quarries on Thasos still preserve numerous grooves created to house the metal wedges used to detach the blocks.

Bottom. Metal wedges and levers are used to detach a large block of white marble at a quarry of Pentelic marble.

Above. Ancient quarrymen digging deep caesuras to detach pieces of rock.

Left and center. Sizes of column drums abandoned at the quarries in Cusa.

Above, right. System of extraction of monolithic column shafts.

Opposite. Detail of the facade of the Celsus Library at Ephesus (early second century A.D.), decorated with columns featuring monolithic shafts and elaborate trabeations.

It was also possible to quarry stone by way of a mine (Gr. *hyponomos* or *dioryx*, Lat. *cuniculus*). This method necessitated deeply piercing the bedrock, often excavating several chambers and branches, thus creating a cave or mine whose upper walls—the "roof" (Gr. *meteoron*) of the cave—were in many instances supported by portions of stone left in place to function as pillars. Among the most interesting examples are the underground mines dug to remove the *lychnites* marble on Paros. Four galleries were dug there, the largest of which, known as the Grotto of the Nymphs, penetrated the mountain for a length of about 130 meters, reaching a depth of roughly 64 meters below the level of the cave mouth. It has been calculated that more than 50,000 cubic meters of the most precious white marble for ancient statuary was extracted from these mines.

Related to mining by caves was the system of mining by wells, employed in Greece from the Archaic period to the Hellenistic, which called for the excavation of relatively small circular wells, from which individual blocks could then be removed, often destined to become the drums of columns. One example of stone mined by a well is the Agrileza marble used in the fifth century B.C. for the construction of two temples at Cape Sounion in Attica.

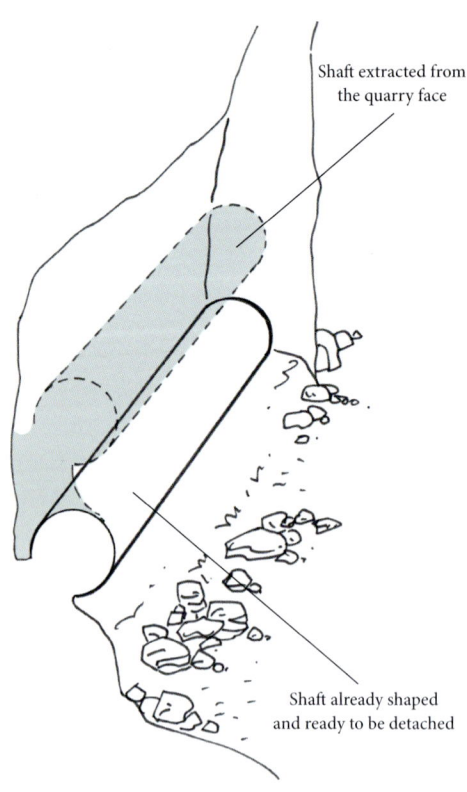

The extraction of a block of stone or marble from a quarry face began by painting or engraving a line on the surface of the bedrock to indicate the shape and size of the piece to be extracted. The quarryman (Gr. *latomos* or *lithotomos*, Lat. *fossor*) then dug along this line with a pick to create a channel (Gr. *aulax*, Lat. *caesura*), the width and depth of which varied according to the ease of cutting the stone; this technique served to isolate on three vertical sides the piece to be extracted, while the upper and front faces were already free because they formed part of the face of the quarry itself. The last phase of extraction involved digging a horizontal groove along the surface of the block to be detached. This groove was shallower than the other channels, and a series of metal wedges (Gr. *sphenes*, Lat. *cunei*) was driven into the bottom of it; following repeated hammering, these wedges would create a long fracture, making it possible to detach the

The elimination of excess material and any imperfections resulting from quarrying took place immediately after detachment of a block from the quarry face. This complex operation required a great degree of specialization on the part of stonecutters.

piece. The wedges were inserted in a row of special holes (Gr. *koila*) created inside the groove; traces of these holes are still present at many ancient quarries, both in soft stone (at the sandstone quarries on the island of Kalidon in Aetolia, for example) and in marble (at the quarries on Thasos and Naxos in Greece, and at those of Belevi in Asia Minor).

Another method of extraction, ingenious in its conception, made use of dry wooden wedges that, once inserted in the holes created within a surface groove, were wrapped in wet cloth; the expansion caused by the wood's absorption of water created progressive pressure, eventually splitting the bedrock. Ancient sources attest to the use of this system as early as the sixth century B.C. at the quarries of white Doliana marble. Other sources, related to stone that was more difficult to extract, refer to the application of fire, immediately followed by cooling with water, which resulted in a thermal shock that split the stone. In all cases the detachment of a piece of stone entailed the use of metal levers and wooden poles.

The construction of colossal temple buildings required the extraction of enormous stone elements, to be used, for example, in the creation of colonnades. This was the case for Temple G at Selinus, a pseudodipteral temple of eight by seventeen columns, begun in the closing quarter of the sixth century B.C. and unfinished when the city was destroyed in 409 B.C. In the area where the stone material was extracted, near Cusa (about 13 kilometers northwest of the ancient city), numerous calcarenite elements remain today that had been destined for the drums of Temple G's gigantic peristyle. The quarry was composed of a large

rocky platform, about 1.3 kilometers long, in which the stone elements were quarried vertically by digging a groove—roughly 0.5 meters wide and more than 5 meters deep—around each piece. Blocks were probably detached from the bedrock with the use of wooden wedges inserted in holes that had been made along the lower level. The future column drum, by now freed from the quarry face, was turned over onto its side and rolled to the quarry yard for rough shaping. The pieces left in the quarry when the work site was abandoned represent various stages of completion. Some were finished and ready for shipment, while others remained attached to the bedrock.

The Romans so perfected quarrying techniques that they could extract monolithic elements of colossal size. In most cases these were column shafts, extracted horizontally by employing methods roughly analogous to those that were used for blocks. Each shaft was cut directly out of the quarry face; while the piece was still anchored to the quarry by only a short section of stone, grooves were cut that almost completely defined the external surface of the future shaft. This method was used to extract the gigantic monolithic shafts in granite that were transported to Rome during the reign of Hadrian for use in the pronaos of the Pantheon and in the Temple of Venus, as well as the shafts in cipollino verde for the Temple of Antoninus and Faustina in the Forum Romanum. The Greeks probably used a similar system for the great temple buildings dating to the Archaic period (including the temples of Apollo at Corinth and those at Syracuse), before the custom of making columns from drums had developed.

Column shafts made for a temple under construction at Kos in the first century A.D. but never finished. Clearly visible are the various stages in dressing the shafts, some already with grooves, others only roughed out.

To produce shafts of smaller dimensions the Romans worked out a method for extracting the pieces in groups (bilobed or quadrilobed), rather than individually, in such a way that the block detached from the quarry contained within it several attached shafts. As reflected in numerous examples deposited at the port of Ostia during the imperial period, this system facilitated transportation and also increased the strength of these long architectural elements, by partially dividing the stone and thus reducing stress from flexion.

To dig out the grooves necessary for detachment of blocks, quarrymen used a scrabbling hammer (Gr. *latomis* or *typis*, Lat. *upupa* or *dolabra*), similar to an axe. This tool had a metal head with a double point (or with a point and a peen) that, particularly in an area of soft stone, left characteristic parallel lines on the rock surface, often with a curving shape. The faces of the quarries at Syracuse, which in the first half of the fourth century B.C. furnished excellent limestone for construction of the city walls, still preserve visible marks left by the blows of a scrabbling hammer.

FROM EXTRACTION TO ROUGH SHAPING: THE CREATION OF UNFINISHED BLOCKS

Once detached, a block was transferred, often by means of wooden rollers (Gr. *phalagges*, Lat. *cylindri*), to a work area—known as the quarry yard—where it underwent rough shaping (Gr. *pelekesis*). This operation was intended to eliminate all the imperfections of extraction, so that the piece was brought to a form near the final shape desired. Executing this step at the quarry saved the effort of transporting excess portions of stone, obviously destined for elimination, to the work site. Elements reserved for making walls in ashlars were given a preliminary squaring, while all the pieces assigned to other parts of the construction (bases, shafts, drums, and capitals) were partially sculpted to obtain an unfinished product, in much the same way as blocks extracted to make sarcophagi and basins.

The rough-shaping phase left a thick residue on the faces of the pieces that had been extracted. Referred to as the protective or sacrificial surface (Gr. *apergon*), this residue protected the final surface of the block during its transportation to the work site.

Various tools were used to execute the rough shaping of stone blocks. Some of these tools employed direct percussion, that is, they were struck directly against stone surfaces. Composed of a metal head and a wooden handle, these tools were quite similar to those still in use today. Included in this group were the cutting hammer (Lat. *doloratorium*), the pick (Gr. *oryx*, Lat. *fossoria dolabra*), and the claw hammer (Gr. *xois charakte tracheia*). Other types of tools depended upon indirect percussion—in other words, they were used together with additional tools—and each had a point or edge that cut into the stone when struck by the second tool. Indirect-percussion tools included the chisel (Gr. *glaris* or *xyster*, Lat. *caelum* or *scalprum*), the gradine (Gr. *xois charakte*), and the punch, which was usually struck by a mallet (Gr. *sphyra* or *kestra*, Lat. *malleus*).

The phase of rough shaping, like that of the extraction that preceded it, was usually related to precise orders from the architect that specified the dimensions of individual pieces. In this regard, an inscription from the second century B.C. concerning the reconstruction of the great dipteral Temple of Apollo at Didyma detailed information about ancient building practices. This engraved report (Gr. *syngraphe*) relates the costs of materials and their working in the quarry, including the number of laborers required: "Work conducted in the quarry under the guidance of Apollas, with 15 men in number, together with 3 assistants. Extraction and rough-shaping of 6 column drums, for a total of 978 cubic feet, and an Ionic capital of 175.2 cubic feet, at 5.5 drachmas per foot, for a total of 6,359 drachmas, 3 chalkus. In addition, 227 small blocks, for a total of 489,75 cubic feet, at 1.66 drachmas a foot: 802 drachmas."

QUARRYMEN AND DIRECTION OF THE QUARRY

The ancient quarry was a place where many people worked, and the complexity of its organization increased according to its size, and consequently the quantity of material extracted. During the Greek period the mining of blocks took place on the basis of contracts between public or private sponsors and a contractor, either the owner of the quarry or a person contracted to oversee the work. Often the contractor was also responsible for the preliminary dressing of stones and their transportation. Ancient sources, in particular inscriptions, relate the duties of various artisans, including simple quarrymen (Gr. *latomoi*), workers in charge of cutting and rough shaping blocks (Gr. *lithourgoi* and *lithotomos* or *lithotomikos*), and experts in the complex modes of transportation (Gr. *lithoulkoi*). Problems related to working materials that were more difficult to cut, such as marble, created a further class of specialized artisans (Gr. *marmaropoioi*). In some cases large masses of prisoners worked in quarries. The historian Diodorus Siculus (*Library* 11.25.23) relates the events at Agrigento (ancient Akragas) after the battle of Himera in 480 B.C.:

> The cities put the captives allotted to them in chains and used them for building their public works. A very great number was received by the Akragantini, who embellished their city and countryside; for so great was the multitude of prisoners at their disposal that many private citizens had five hundred captives in their homes. . . . Most of these were handed over to the state, and it was these men who quarried the stones of which not only the largest temples of the gods were constructed but also the underground conduits were built to lead off the waters from the city.

During the Roman period the organization of quarries and the control of materials extracted were highly structured, perhaps on the model already elaborated by the Ptolemies in Egypt. Suetonius (*Tiberius* 49.2) relates that Tiberius took from a great many cities the right to exploit mines and quarries of stone by means of an act of expropriation that made the quarries, especially those for Luna marble, among the properties of the emperor (Lat. *patrimonium Caesaris*). This policy, probably adopted previously under Augustus, was

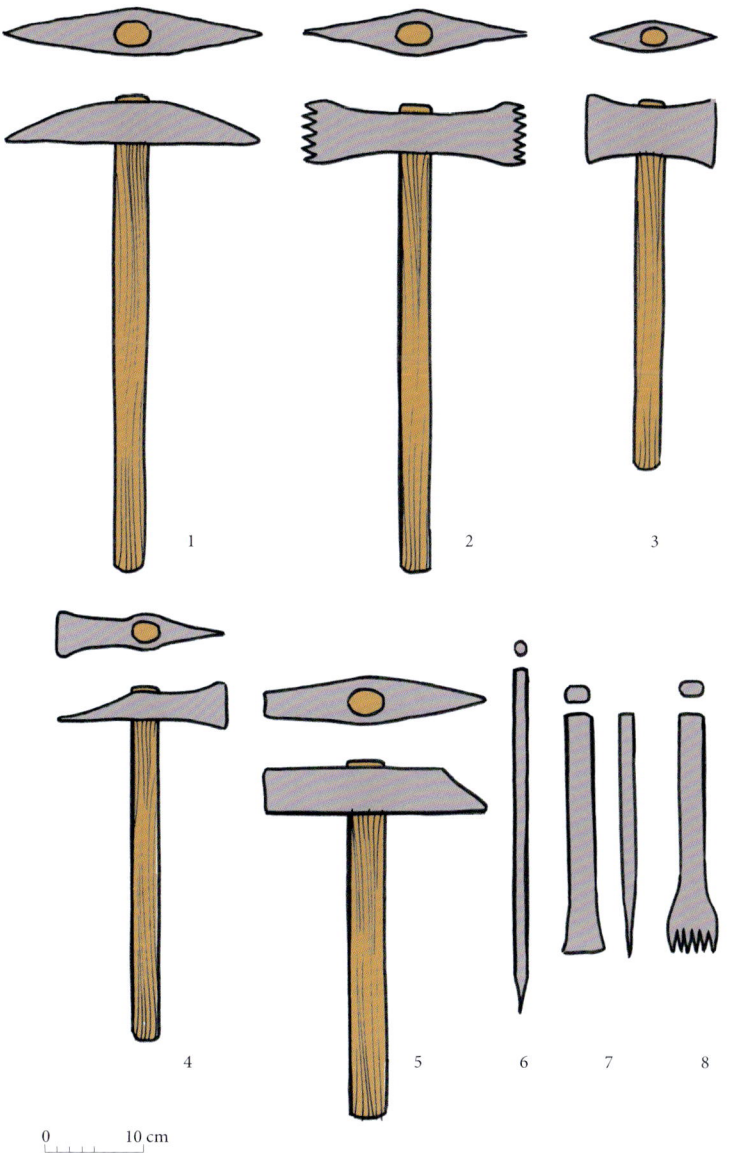

The primary direct- and indirect-percussion tools used for working stone materials: 1. Scrabbling hammer; 2. Claw hammer; 3. Cutting hammer; 4. Stonemason's hammer, or kivel; 5. Mallet; 6. Punch; 7. Chisel; 8. Claw chisel.

continued by Claudius, as indicated by the name Mons Claudianus given to the Egyptian site of extraction of the stone known as granito del foro. Beginning with the rule of Antonius Pius, quarries began the prefabrication of construction materials without specific orders.

At the top of the system that oversaw the activities of extracting and distributing marble was the *statio marmorum*, a special office directed by a *procurator marmorum*; beneath it were the various *procuratores* who, in turn, administered individual quarries. There were additional officials of various levels (Lat. *tabularii a marmoribus*), as well as accounting officials (Lat. *dispensatores*). Finally, certain inscriptions indicate the existence of overseers for single types of marble, such as the inspector of quarries of cipollino verde (Lat. *a lapicidinis Carystiis*) on Euboea and the official legate for administration of Luna marble (Lat. *tabularius marmorum lunensium*).

Those pieces not made for public buildings (Lat. *ratio urbica*), and those that were not imperial property (Lat. *ratio imperialis*), were offered by the state to merchants (Lat. *negotiatores*) and to individuals in charge of supplying marble to workshops and construction worksites (Lat. *redemptores marmorarii*).

Some architectural pieces were made for specific buildings or cities. An inscription on a block of cipollino marble found in Rome and dated to the Hadrianic period relates that it was made for the emperor's buildings (Lat. *ratio domi Augusti*), while another inscription, on a block of black marble, indicates that the piece was sent directly from the quarry to the *splendidissima colonia* of Leptis Magna, in northern Africa. Far more frequent was the case of inscriptions put on blocks or shafts at the moment of their extraction or on their arrival at the *statio marmorum*. These inscriptions generally included the consular date and, in the case of blocks recently extracted at a quarry, the term *loco* followed by a number that signified the wall of the quarry and thus identified the site of extraction. This information was often accompanied by other facts, such as the name of the workshop that had dressed the material or even that of the commissioner (Lat. *probator*) in charge of examining the quality of pieces before shipment. Other numbers indicated the quantity of blocks mined. An inscription on a block of cipollino verde, for example, reports that it was number 2,400 in the sequence of pieces quarried.

Until at least the middle of the first century A.D., Rome's *statio marmorum* was located on the banks of the Tiber, in the area of Marmorata at the foot of the Aventine. Several hundred blocks of white and colored marble have been found there. This site was the distribution center for all of the city's marble building elements, which arrived on boats by way of the Tiber from the port of Ostia. Later, perhaps due to increased requests related to imperial and public architecture, as well as those commissions from private persons, it was decided to create an area with greater capacity, one that could accommodate both larger shipments and additional types of commodities. The area chosen was the Portus Ostiensis, where special storerooms were constructed—first around the port of Claudius and then along the wharves of the even larger port of Trajan—to hold marble until it could be shipped to the numerous work sites in Rome. Blocks were brought there by large boats (Lat. *naves lapidariae*) that arrived from various ports throughout the Mediterranean. Every area with a quarry was assigned a port located in the nearest city. Materials quarried in Egypt were loaded at the great port of Alexandria; violet granite arrived following maritime routes from Alexandria Troas; the marbles of Attica were shipped from Piraeus; and Ephesus was the source of white marble extracted from quarries near that city, as well as pavonazzetto from the Docimian quarries in Phrygia. Similarly, Nicomedia was the center of massive exportation of Proconessium marble and the so-called peacock's eye, as was Carthage for many marbles quarried in northwest Africa.

CLAY AND TERRACOTTA

Men, in the old way, were born like animals in forests and caves and woods, and passed their life feeding on the food of the fields. Meanwhile, once upon a time, in a certain place, trees, thickly crowded, tossed by storms and winds and rubbing their branches together, kindled a fire. Terrified by the raging flame, those who were about that place were put to flight. Afterward when the thing was quieted down, approaching nearer they perceived that the advantage was great for their bodies from the heat of the fire. They added fuel, and thus keeping it up, they brought others. . . . Therefore, because of the discovery of fire, there arose at the beginning, concourse among men, deliberation, and a life in common.

From On Architecture *(2.1.1) by Vitruvius, the illustration that accompanies his passage on the origins of architecture, in the edition published in 1521 by Cesare Cesariano at Como, Italy.*

Opposite. Ancient Corinth arose at the center of the large gulf that took its name, and grew in importance beginning in the eighth century B.C., when it became a flourishing center for artisan crafts, from metalwork to ceramics. Visible here are excavations of the city against a background of the impressive Acrocorinth, the mountain that was also the acropolis of the polis.

... They began, some to make shelters of leaves, some to dig caves under the hills, some to make of mud and wattles places for shelter, imitating the nests of swallows and their methods of building. Then observing the houses of others and adding to their ideas new things from day to day, they produced better kinds of huts.... And first, with upright forked props and twigs put between, they wove their walls. Others made walls, drying moistened clods which they bound with wood, and covered with reeds and leafage, so as to escape the rain and heat. When in wintertime the roofs could not withstand the rains, they made ridges, and smearing clay down the sloping roofs, they drew off the rainwater.

With this highly evocative passage from the first century B.C., Vitruvius (2.1.1–3), illustrates the beginning of architecture, locating its place within the more general process of human civilization. Together with his reference to shelters made in caves, the Roman architect connects the first steps of building to the use of material made available by nature: reeds, branches and wood from trees, earth, and mud. Archaeology has demonstrated that the earliest materials used in simple forms of architecture were in fact wood and clay—the first because of its great availability, the second for the plastic and waterproofing properties that make it highly versatile—followed quickly by additional applications that made use of highly ingenious technical solutions.

CLAY

Clay was the protagonist of one of the most important discoveries of the human experience. Indeed, it was sometime in prehistory when humans learned not only that an object made of damp clay maintained its shape when dried, but that it became hard when baked.

Clay (Gr. *argilos*, Lat. *argilla*) results from the slow breakdown and decomposition of feldspathic rock, through the effects of atmospheric agents, into particles with a diameter of less than one-sixteenth of a millimeter. Sedimentary, incoherent, and with an earthy appearance, feldspathic rock is principally composed of hydrous aluminum silicate. The compost that forms this argillaceous substance has a colloidal nature that confers on clay the capacity to absorb water in a quantity equivalent to as much as 70 percent of its volume, transforming it into a highly malleable plastic mass. Through the process of drying, clay releases part of the absorbed water, but only with firing at a high temperature does it lose all the water present, allowing it to maintain the shape it had been given in the modeling phase. The color of clay varies across a spectrum that runs from white to dark brown and includes red and gray-green; the specific color depends on the individual iron compounds that are present and on the amount of carbon substances with an organic origin.

Pure clay, without the addition of water, does not possess the qualities required for modeling, and it cracks during the drying phase. Thus it is necessary to combine clay with aggregates—sand and plant material may be used, as well as ground terracotta—that function as "degreasing" agents. Only the so-called lean clays, those found naturally mixed with sand, can be worked with the addition of water alone.

Greece, Asia Minor, and the Greek cities of the West, as well as Rome and many provinces of the empire, were all rich in deposits of good-quality clay. Using the raw material extracted from quarries located at those sites, ancient artisans created numerous types of objects, ranging from construction material to crockery for banquets, and from containers for the transportation of foodstuffs to sculptures and sarcophagi. Corinth and the surrounding region were among the areas providing clay of superior quality, and the city itself became one of the principal areas

Corinth was the source of a series of painted terracotta panels depicting scenes related to working clay. This panel, today preserved in Berlin, depicts the operations of quarrying clay and is dated around 550 B.C.

of ceramic production in the Greek world. In Corinth alone three different types of clay have been identified: two of these, one red in color and the other white, are present on Acrocorinth, the city's ancient acropolis; the third, light in color, is found in the area of the plain. Excavations have uncovered within the city one of the artisan quarters that was involved in working clay. Located behind the walls near the western end of the urban area archaeologists have found the wall structures of several workshops (Gr. *ergasteria*), along with a large quantity of ceramic products (including vases and statuettes) and more than a hundred molds used for their creation.

THE CYCLE OF WORKING CLAY

As with stone, the site of clay extraction was a quarry, which according to the specific geological formation of an area could be exploited in one of two ways: either by digging an open-pit system or by excavating mines. Following identification of the layer of clay suitable for use in mixtures, a trench was created that progressively assumed the shape of steps. A series of painted terracotta tablets (Gr. *pinakes*) from Corinth depicts scenes related to working clay. One of these tablets, dated around 550 B.C. and today preserved in Berlin, shows a scene of clay quarrying: on the right, with the help of a pick, one figure is busy extracting material from a quarry face; in the center a second figure leans down to collect clay in a basket; and on the left a third man passes his basket of clay to a fourth coworker.

After extraction and a first period of drying in the sun, clay was aged. During this phase the process of oxidation of pyrites in the material led to the formation of sulfates; the latter are soluble in water and could easily be washed away by rain. The material was next purified of all extraneous substances that might be harmful to the later phases of modeling and firing. Both Greek and Roman artisans attained very high levels of achievement in the techniques of purifying clay. The system was based on the principles of suspension and sedimentation: Clay was mixed with an abundant quantity of water in a tub; after a time the heavier substances began to deposit on the bottom and, in the uppermost area of the water, a very fine argillaceous suspension, known as slip, formed and was collected by the potter with the use of a bowl. This method of purification could be further refined using running water and a series of stacked tubs of various sizes. Clay was put in the topmost container and initially purified by sedimentation; the slip produced then overflowed into the lower container, to undergo another phase of purification; and so on. After the slip was collected from the final tub, it was dried; the clay that resulted could be used immediately by the potter for modeling products, or it could be stored in a dry place.

RAW CLAY AND THE CONSTRUCTION TECHNIQUE OF *PISÉ*

To make the majority of products destined to be fired, quarried clay required long processes of aging and purification, but all of these steps

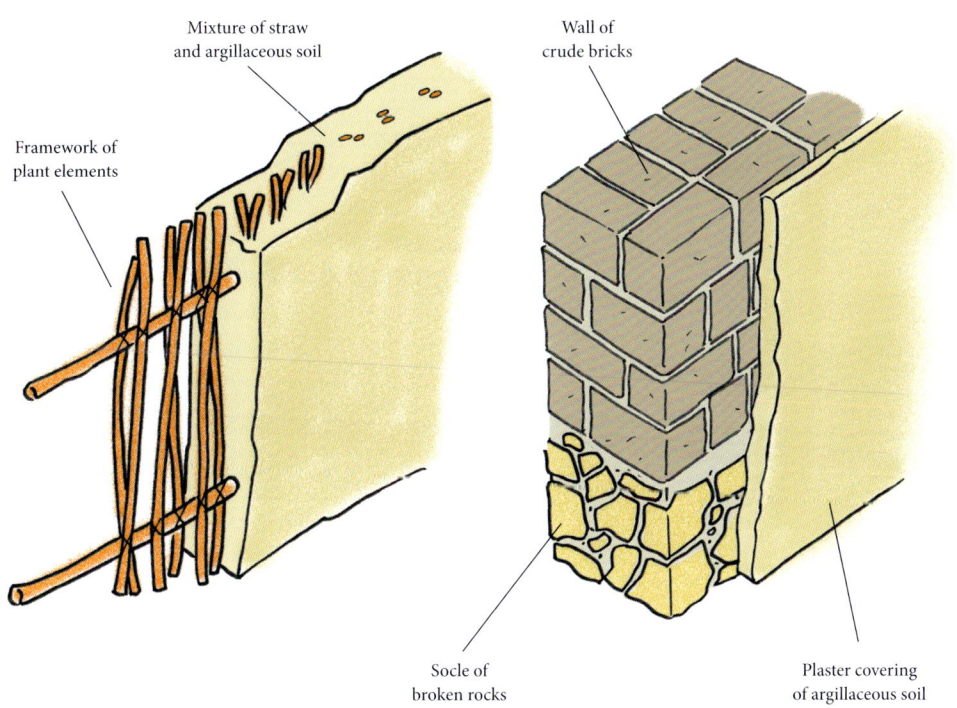

Construction techniques using perishable material, including wood and clay: at left, creation of a wall using a framework covered with an argillaceous mixture; at right, wall with a socle in stones bound by soil.

could be greatly simplified for construction techniques that made use of raw clay. In terms of architecture the builder's concern was not for the purity of the basic material—a matter of great importance in the case of the most refined ceramics—but instead for the nature and percentage of degreasing aggregates, which served to give body and strength to the mixture. Clay for construction was placed in pits that were filled with water and then mixed by foot with degreasing agents of both plant (straw, chaff or husks, and dried grass) and mineral (sand and gravel) origin. As noted above, lean clays—those found in nature already mixed with a degreasing agent—do not require the addition of anything other than sufficient water to make the clay mass plastic. According to some ancient sources, there were two different types of mixtures used for construction: the first, known by the French term *pisé*, was based on lean clay mixed with aggregates of mineral origin; the second (Gr. *pēlos echyrōmenos*, Lat. *lutum paleatum*), referred to as *torchis* (from the Lat. *torquere*, meaning "twist"), was characterized by the addition of plant elements such as straw, which had to be twisted to be cut.

As discussed, materials used in the most ancient architecture were those directly available in nature, especially clay, wood, and other plant elements. These were skillfully combined in ways that would best exploit their respective physical properties and static capacities. The basic construction system called for creation of a wooden skeleton, to which was anchored a trellis in plant material, in turn covered with a mixture of clay, straw, and sand. Vitruvius (2.1.5) recalls the use of this technique among certain Anatolian peoples: "The Phrygians, who are dwellers in the plains . . . fasten logs together at the upper end, and so make pyramids. These they cover with reeds and brushwood and pile up very large hillocks from the ground above

CLAY AND TERRACOTTA 45

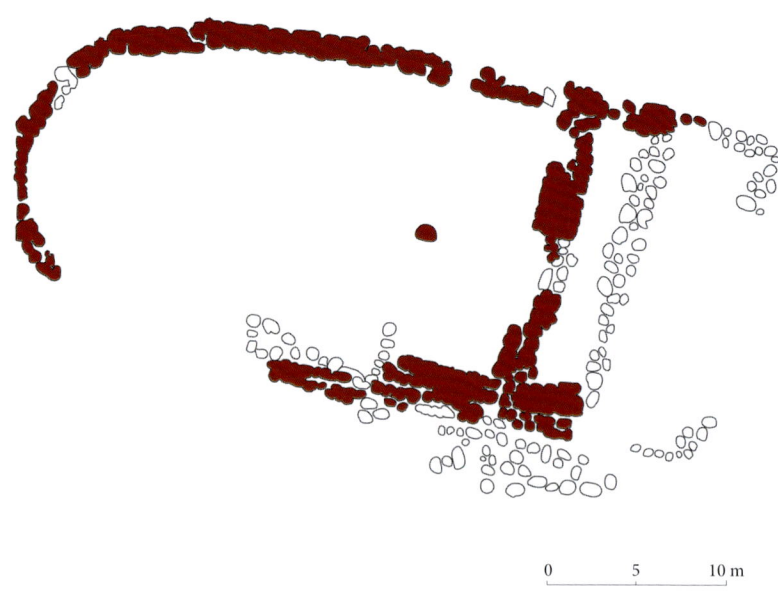

Layout of one of the apsidal houses on the site of Nichoria in Messenia.

their dwellings. This arrangement of their dwellings makes the winter quite warm, and the summer cool." Unfortunately, the highly perishable nature of these materials makes it difficult to identify them during excavation. In the most fortunate archaeological contexts ancient buildings were destroyed by fire, and as a result the argillaceous mixture used for the walls of the structure was indirectly baked, and retains the impression of the trellis that it covered. Construction by means of a trellis has been revealed in various Greek structures that date to the Protogeometric period, as for example in a structure built during the tenth century B.C. at Nichoria in Messenia. Over the course of the Geometric period, the system was largely adopted for the construction of houses with an apsidal shape, since it better facilitated the building of curvilinear rear walls than did other building methods.

Research has demonstrated that the trellis technique, because of its rapid execution and use of material commonly available throughout the Mediterranean, was widely employed not only in Greece but also in the Etruscan and Roman world. At Spina, an Etruscan city on the Po plain, houses were built with large wooden posts serving a support function, joined together by horizontal beams. The result was a structural framework whose squares were filled with interwoven reeds and branches; the latter were covered with an argillaceous mixture, which was smoothed and slightly baked to make it waterproof and enable it to serve as insulation. Even earlier examples of the use of similar frameworks exist in Rome. On the southwestern summit of the Palatine have been found the traces of several huts dated to the eighth century B.C., during the period of the mythical foundation of the city by Romulus. One of these structures, measuring 4.9 by 3.6 meters, was built by inserting a series of seven vertical poles (with two more near the front portico) in the rocky soil; these seven poles supported walls of clay and straw as well as a roof made of plant material.

Following development of other construction techniques, the trellis method—although it too had evolved—was replaced by systems that were more durable and offered greater static capacities. It remained in use only for internal partitioning and for building minor structures. Vitruvius noted his dissatisfaction with the trellis method: "I could wish that walls of wattlework had not been invented. For however advantageous they are in speed of erection and for increase of space, to that extent are they a public misfortune, because they are like torches ready for kindling. . . . These also make cracks in the plaster covering owing to the arrangement of the uprights and crosspieces. For when the plaster is applied, they take up the moisture and swell, then when they dry they contract, and so are rendered thin, and break the solidity of the plaster."

Although based on the use of similar clay-based mixtures, another construction method noted by Pliny the Elder (*N.H.* 35.169) involved a completely different work-site procedure. He praises "walls made of earth that are called framed walls (Lat. *formaceos*), because they are made by packing in a frame enclosed between two boards, one on each side, and so are stuffed in rather than built," and asks, rhetorically, "do they not last for ages, undamaged by rain, wind, and fire, and stronger than any quarry-stone?" The reference is to a technique widely adopted in ancient architecture based on the use of wooden formworks (Lat. *formae*) that served to establish the width of a wall. These were erected by mounting upright pilings (Lat. *tabulae*), usually fixed by way of crossbeams or ties (Lat. *catenae*). The clay mixture was carefully pressed down into the formwork with a heavy wooden mallet (Lat. *fistuca*), to compact it and eliminate at least some of the water. Construction of a wall of this type was executed in sections of limited height and width; once the mixture had dried and solidified, the mason removed the formwork and set it up at the appropriate spot to make the next wall segment.

Pliny introduced the passage on *formacei* walls with a precise reference to their African and Iberian context, and in fact the best examples of walls made using this technique are preserved in Spain. At Ampurias, the city founded by Greeks at the midpoint of the coast along the Gulf of Rosas, several houses built between the first century B.C. and the first century A.D. still retain fairly tall structures of this type. Their walls are 50 centimeters thick, and clearly visible on their surfaces are impressions left by wooden formworks and the poles used to attach them.

PRODUCTION OF MUD BRICKS

Yet another use of an argillaceous mixture, by means of a very ancient method with roots in Greece that reach back to the sixth and fifth millennia B.C., is represented by the mud bricks (Gr. *plinthos ome*, Lat. *later crudus*) employed in Greek and Roman architecture. In fact, the first evidence for the use of mud bricks is found at the Neolithic settlement of Sesklo in Thessaly, which dates to that period.

This construction system—still employed, more or less unchanged, in some parts of the world—is based on a mixture of argillaceous

Wooden formwork for shaping bricks

Top. Shaping crude bricks.

Bottom. Method for making a wall of the type defined by Pliny the Elder (N.H. 35.169) as formaceo.

CLAY AND TERRACOTTA

Modern construction using crude bricks, in the heart of Thrace. Note the high socle in stone, necessary to keep damp ground from damaging the walls.

soil, straw, and water that is then shaped into the form of a parallelepiped with the use of a bottomless mold (Gr. *plaision*, Lat. *forma*). The importance of straw in such mixtures has been revealed through a well-known biblical passage from Exodus (5:7–8) that relates to the punishment of the Jews by the pharaoh Thutmose II (or Ramses II): "Ye shall no more give the people straw to make brick, as heretofore: let them go and gather straw for themselves. And the tale [quantity] of the bricks, which they did make heretofore, ye shall lay upon them; ye shall not diminish ought thereof." Indeed it was from the Egyptian world, in particular from the decorations of several tombs found there, that valuable iconographic documents relating to the working of clay have originated. In two different tombs at Thebes there are depictions of mixing clay by treading on it, as well as illustrations of the shaping and drying of bricks.

Vitruvius (2.3.1) emphasizes the significant role played by the choice of raw material to ensure good-quality products resistant to the effects of time: "First I will speak about bricks, and from what kind of clay they ought to be brought. For they ought not to be made from sandy nor chalky soil nor gravelly soil: because when they are got from these formations, first they become heavy, then, when they are moistened by rain showers in the walls, they come apart and are dissolved. And the straw does not stick in them because of their roughness. But bricks are to be made of white clayey earth or of red earth, or even of rough gravel. For these kinds, because of their smoothness, are durable. They are not heavy in working, and are easily built up together."

Also important to the quality of construction were the specific times for drying bricks in

View of the Sanctuary of Demeter at Eleusis.

the sun, as he relates (2.3.2): "Bricks are to be made either in the spring or autumn, that they may dry at one and the same time. For those which are prepared at the summer solstice become faulty for this reason: when the sun is keen and overbakes the top skin, it makes it seem dry, while the interior of the brick is not dried. And when afterward it is contracted by drying, it breaks up what was previously dried. Thus bricks crack and are rendered weak. But, most especially, they will be more fit for use if they are made two years before. For they cannot dry throughout before."

Failure to observe these precautions would have grave effects on the functioning of the wall structure: "When they are built in fresh and not dry, and the plaster is put on and becomes rigid, they remain solid only on the surface. Hence they settle and cannot keep the same height as the plaster. For by contraction and the consequent movement they cease to stick to the plaster, and are separated from their union with it. Therefore the wall surfaces are separated from the wall itself, and because of their thinness cannot stand of themselves and are broken, and the walls settling haphazard, become faulty."

To all these problems discussed by Vitruvius—each of which is intimately tied to the standard cycle of mud-brick production—was added yet another, shared by all the building techniques based on the use of unfired clay: the material's low resistance to dampness from the ground, a quality related to the high-absorption capacities of clay itself. However, at a relatively early stage in the ancient world, this problem was resolved by making, at the bottom of walls, a socle composed of river rocks or stone shards that were collected directly on the construction site.

CLAY AND TERRACOTTA 49

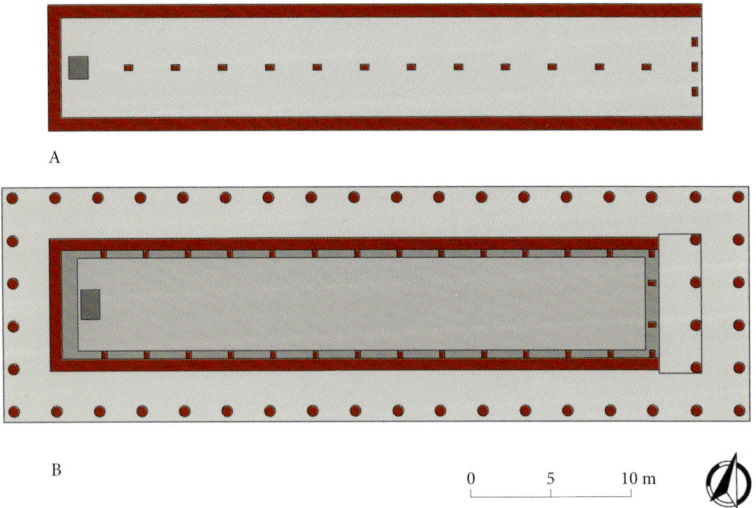

Top. Layouts of the most ancient stages of the Temple of Hera at the sanctuary on Samos: A. First half of the eighth century B.C.; B. First half of the seventh century B.C.

Opposite, bottom. Layout of the building constructed in the ninth century B.C. at Lefkandi in Euboea.
1. *Portico*
2. *Vestibule*
3. *Main area*
4. *Burials*
5. *Central colonnade in the main area*
6. *Minor spaces separated by a hall*
7. *Apsidal thalamos*
8. *External colonnade*

MUD BRICKS IN GREEK AND ROMAN ARCHITECTURE

Vitruvius (2.3.3), followed by Pliny the Elder (*N.H.* 35.171), lists the various types of bricks produced in the Greek world and gives their individual names: "There are three kinds of bricks: one which in Greek is called *lydion*, that is the one which we use, a foot and a half long, a foot wide. Greek buildings are constructed with the other two. Of these, one is called *pentadoron*, the other *tetradoron*. Now the Greeks call the palm *doron*, because the giving of gifts is called *doron*, and this is always done by means of the palm of the hand. Thus the brick that is of five palms every way is called *pentadoron*; of four palms, *tetradoron*. Public buildings are erected with the former; private buildings with the latter." The unit of measure of the palm (Gr. *palaiste*, Lat. *palmus*) corresponds to one-quarter of the base unit, the foot (Gr. *poys*, Lat. *pes*), using the Attic-Cycladian foot of 29.6 centimeters, equivalent to the size of a Roman foot. Thus a *lydion* brick measured 29.6 by 44.4 centimeters; a *pentadoron*, 37 by 37 centimeters; and a *tetradoron*, 29.6 by 29.6 centimeters. These forms are well documented in archaeological excavations, including those in Athens, Apollonia (Epirus), and the habitations of Heraklea Lyncestis in Macedonia. There was also the *hēmiplinthos*, created by dividing a brick in half, and the *triemipodios*, with a square shape and sides equal to a foot and a half (44.4 by 44.4 centimeters). Its use is known, for example, in the Athenian walls of the fourth century B.C. (45 by 45 centimeters, 8 centimeters thick) and in those at Corinth erected at the end of the same century (45 by 45 centimeters, 9 centimeters thick). In addition, excavations have demonstrated that many other types of unfired bricks also existed in the Greek world, with sizes that varied according to the system of measurement employed. For example, there are the bricks used in the fortifications of ancient Smyrna between the ninth and eighth centuries B.C. (each brick measured about 50 by 30 centimeters, and was 8 to 13 centimeters thick); those used in construction of the sanctuary at Eleusis in the fourth century B.C. (49.2 by 49.2 centimeters, about 9 centimeters thick); and those produced also in the fourth century at Gela in Sicily for construction of the city's walls (40 by 40 centimeters, 7 to 8 centimeters thick). These examples testify to the widespread use of mud brick for the construction of defensive walls, usually built on top of a socle with a stone foundation. In Athens an inscription relating to the restoration of the city's walls in 307–306 B.C. tells us that horizontal and crosswise wooden beams were inserted to reinforce the mud-brick walls.

The use of mud bricks characterized the great Greek architecture made from the end of the eighth through the seventh century B.C. The walls of the Temple of Hera at Samos (phase 1, eighth century B.C.; phase 2, about 660 B.C.) were made of mud bricks, as were those of the temple found at Ano Mazaraki in Achaea (late eighth–early seventh century B.C.), the Temple of Apollo at Corinth (675–650 B.C.), and the Temple of Poseidon at Isthmia (also 675–650 B.C.). Even when builders began to construct religious architecture in forms based on stone, the ancient technique based on the use of clay mixtures dried in the sun continued to be widely used at least until the Hellenistic period. This preference was due to the great ease of making such bricks, and thus their low cost, as well as to the abundance of quarries for good-quality clay found in many areas of the Mediterranean—in contrast to building stone, which sometimes required transport across great distances to reach a work site. The mud-brick technique was commonly used in private, architecturally modest structures (such as the houses at Olynthus from the fourth century B.C.), but also appears in buildings of a certain prestige. Outstanding examples from the Hellenistic period include the Pompeion of Athens, made around 400 B.C.; the palaestra of the Sanctuary of Zeus at Olympia; some parts of the gymnasium of the Sanctuary of Asclepius at Epidaurus; and the Temple of Apollo at Thermos in Aetolia, in its reconstruction of the third century B.C. To these can be added several structures noted by Vitruvius (2.8.9–10): "In some cities we may see both public works and private houses and even palaces built of brick. . . . At Tralles there is a palace built for the Attalid kings. . . . There is the palace of Croesus, which the people of Sardis dedicated to their fellow citizens for repose in the leisure of their age. . . . At Halicarnassus also, although the palace of the mighty king Mausolus had all parts finished with Proconnesian marble, it has walls built of brick."

There is ample evidence for the use of mud bricks in the Etruscan and Roman worlds as well. Archaeological research has shown the presence of mud bricks in one of the oldest houses discovered at Acquarossa, dated to the second half of the seventh century B.C. At Roselle mud bricks characterized numerous work sites from the Archaic period, and the same material was also used there for construction of the city's walls, which were built on a foundation made of stone blocks. Other bricks have been found in homes built in the fifth century B.C. at the centers of Pyrgi, Vetulonia, and Marzabotto (the latter is known only by its modern name).

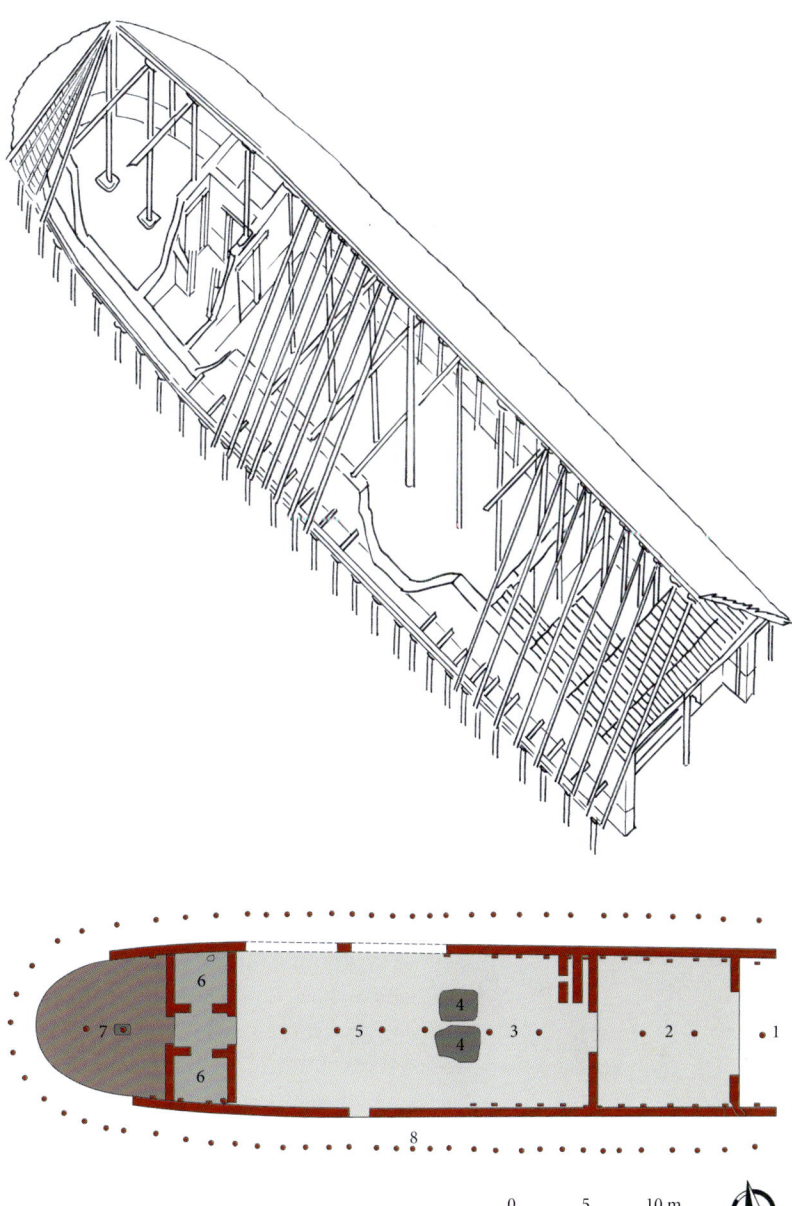

THE HEROÖN OF LEFKANDI AND THE DAPHNEPHOREION OF ERETRIA: CONSTRUCTION TECHNIQUES OF WOOD AND CLAY

Among the most important and best-known apsidal buildings from the Protogeometric period is a monumental structure located near the small modern village of Lefkandi, on the island of Euboea. Made during the first half of

Top. The apsidal building at Lefkandi constitutes the most monumental example of Protogeometric architecture. This reconstruction shows the large roof in wood and plant material, which extended beyond the walls of the structure to form a long external colonnade.

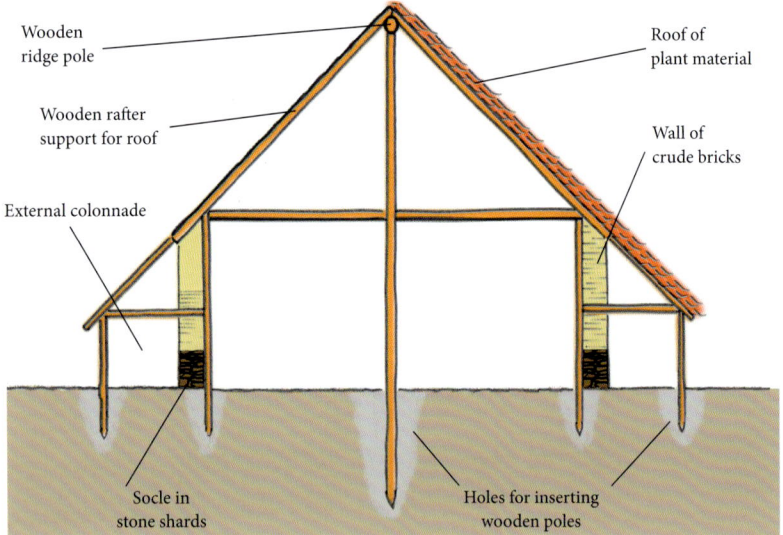

Reconstructed cross section of the building at Lefkandi, with an indication of the building system.

Opposite, bottom. The Sanctuary of Apollo Daphnephoros ("Laurel Bearer") at Eretria presents a complex stratification of cult buildings, including 1. Walls from the Geometric period, and 2. the Daphnephoreion. Also indicated are remains from Temple D (ca. 700 B.C.), with axial colonnade; a late-Archaic temple (520–490 B.C.) destroyed by the Persians in 490 B.C.; a Doric peripteral temple (480–460 B.C.); and an altar-bothros.

the tenth century B.C., it reached an impressive size (approximately 10 by 47 meters), with an elongated form and a large rear apse. The most signicant aspect of the discovery was a burial mound located inside the central room, containing a man and a woman, as well as four horses. The entire building was intentionally destroyed a few decades after its construction, and a large circular tumulus was created over its remains. The site thus became a *heroön*, or hero's grave, the location of cult practices for the dead warrior hero, whose ashes had been deposited in a beautiful bronze crater made on Cyprus and dated to the twelfth century B.C.

The remains of the structure have provided a great deal of technical information about its construction. The building's outer walls stood on top of a socle roughly 60 centimeters wide, constructed directly on the bedrock. Made of a local grayish marble, the socle had what are called double-faced walls, the space between the walls filled with earth. The walls that rose above the socle were made of mud bricks—short stretches still preserve them in situ. Some of the structure's floors were made of beaten clay on top of a loose foundation of rubble that allowed drainage; the inside walls were covered with a thin layer of plaster.

The thresholds located at the openings in the building are evocative, for they are made of wood, recalling the thresholds of ash in the mythical palace of Odysseus at Ithaca.

Especially noteworthy was the presence, along the wall's entire internal facade, of numerous wooden pillars, each with a rectangular section that measured about 10 by 20 centimeters, located at intervals varying from 0.80 to 2.15 meters. Corresponding to each of these pillars was an analogous support located on the exterior of the construction, at a distance of about 1.8 meters from the wall. Another series of wooden poles, about 40 centimeters in diameter, ran along the central axis of the building, with a distance of roughly 3 meters between the centers of the poles. Postholes were dug into the rock outside the building; according to the excavators' calculations, the size of these holes (maximum width 1.45 meters; depth 1.40 meters) would have permitted the erection of wooden posts as high as 8.50 meters. On the basis of this information it has been possible to reconstruct the complex system of the building's roof, which apparently had two slopes and a half-cone termination at the apse. It is probable that the structure included a sort of attic, created with a wooden floor at the terminal height of the walls, and extending to the height of the imposts for the slopes of the roof. The beams of the roof ridge rested on axial uprights, while the oblique rafters unloaded their weight on the perimeter walls, which were reinforced at the contact points by wooden posts. Externally, there was a sort of continuous portico (perhaps an upper gallery) formed by the roof, which extended to reach the external wooden posts. The roof itself was made of interwoven reeds and rushes, carbonized remains of which were found on the floor in the main room.

A large temple dedicated to the worship of Apollo Daphnephoros ("Laurel Bearer") was built at Eretria between the sixth and fifth centuries B.C. Excavations beneath its structure have brought to light the remains of an earlier temple, an apsidal hekatompedon (literally, "one hundred feet"), that is about 37 meters in length. Other oval constructions from the Geometric period have been identified alongside this structure, as well as the remains of an older building with an apsidal shape, dating to the first half of the eighth century B.C. Measuring 6.50 by 9.75 meters, this earlier structure had a short anterior portico that was supported by two wooden columns positioned almost in line with the side walls. The construction technique involved a masonry socle extending 40 centimeters in height and built with irregularly shaped stones. The entrance door, open between the two rectilinear walls and slightly raised on a section of masonry, was defined by the presence of two clay bases (each 42 centimeters wide), which perhaps functioned as rests for the doorjambs or for wooden support elements.

The absence of solid archaeological evidence related to the upper part of the building's walls made it difficult to identify the architectural system used. Archaeologists have

Top. Remains of the apsidal building identified with the Daphnephoreion of Apollo. At left runs the cella wall of the late-Archaic-period Temple of Apollo.

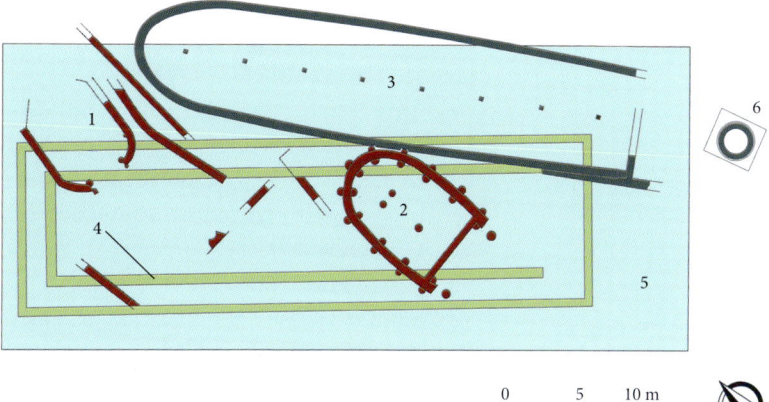

Hypothetical reconstruction of the roofing system of the Daphnephoreion at Eretria.

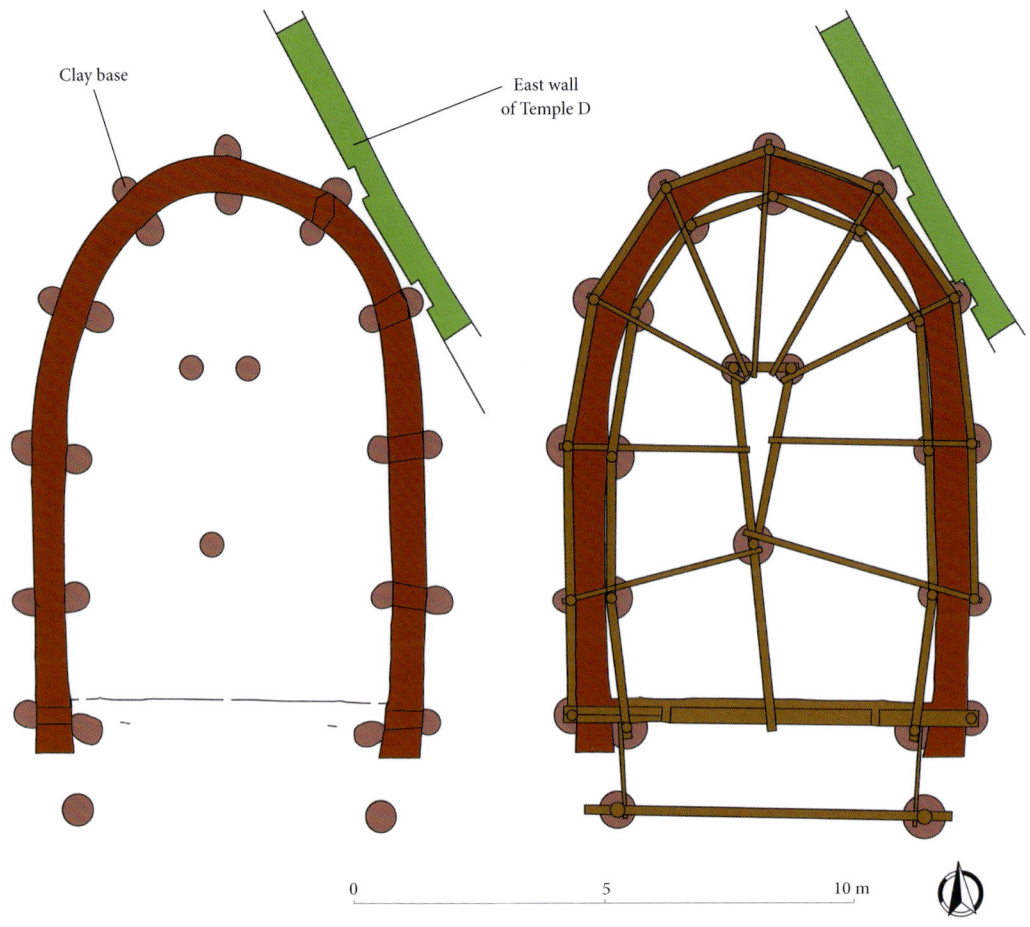

hypothesized that the construction reproduced the Daphnephoreion of the sanctuary at Delphi, the mythic temple of laurel boughs built by Apollo himself. If this theory is correct, it would suggest that the Greek imagination associated the natural element of laurel leaves, an attribute of the god, with an architectural practice based on the use of perishable material. This Delphic example would not have been an isolated case, for various traditions tell of a temple made of willow branches in the Sanctuary of Hera on Samos, and of another made of beeswax in the Sanctuary of Apollo on Delos. However, recent studies based on a reexamination of the excavation results at Eretria have more closely identified the construction's architectural character, and raised numerous doubts about the likelihood that the temple was made of laurel branches. The perimeter walls were most likely built with a mixture of argillaceous earth left to dry in the sun, or with traditional mud bricks.

While the building technique of the temple at Eretria may reflect the construction tradition based on the use of a stone socle and walls made of clay, the insertion of wooden supports beside the walls constitutes a significant, innovative aspect. Along the exterior of each of the building's side walls, archaeological excavations have identified a series of circular bases in clay, each with a diameter of about 45 centimeters; another three bases of the same type were found inside the construction, arranged to correspond to the points of a triangle almost axial with the monument. These clay elements served as supports for the wooden

posts reinforcing the walls, and as supports for the roof, although we do not know if they did so as true bases, or as fillings for the post-holes to protect the wood from the dampness of the ground. Based on the positions of these bases, it is possible to reconstruct the system of the structure's roof, which appears to have been made with a group of rafters extending from three ridge poles and arranged in a fan shape; the poles formed a characteristic triangular space at the center, perhaps an opening for the exit of smoke from the inner hearth. On the outside of the structure the oblique poles rested on wooden pillars that flanked the wall, and were connected above by a double series of small beams. The triangle formed by the two inclined slopes of the roof and the horizontal beam of the rectilinear entrance wall—the area destined to become the tympanum of a stone temple—probably contained a window, aligned with the doorjambs of the door beneath it, to provide illumination for the interior.

FIRED BRICKS

During the fourth century B.C. the Greeks began to bake mud bricks in kilns. In the beginning such material was used exclusively for domestic architecture, as indicated by homes in Ionian Abdera and Thrace, and on Thasos; but at Kassope in Epirus, in the second half of the fourth century B.C., baked bricks were used not only for homes but also for the Katagogion, a large building erected near the agora. The use of baked bricks also spread rapidly in Magna Graecia, with numerous

The monumental brick structure of the Baths of Diocletian in Rome. The largest bathing complex in Rome, it was built in just eight years, from A.D. 298 to 306.

Below. Open-air kiln and furnaces for baking clay.

Opposite. View of the great brick structure of Trajan's Market in Rome. Note the flat and relieving arches inserted in the walls to support openings.

examples at Locri and Velia, and in particular ancient Rhegium (modern Reggio Calabria), where this material also served for the construction of many chamber tombs covered with a barrel vault, also made of brick. Nevertheless, it was not until Rome's imperial period that the production and use of material made of baked clay occupied the dominant position in all fields of construction. The Basilica of Pompeii, erected in the second half of the second century B.C., reveals that even during the late republic the Romans were skilled at the

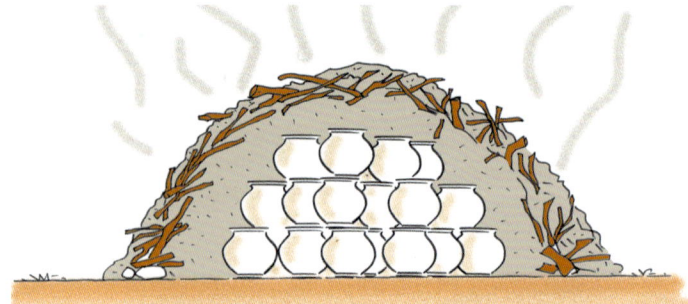

Open-air kiln

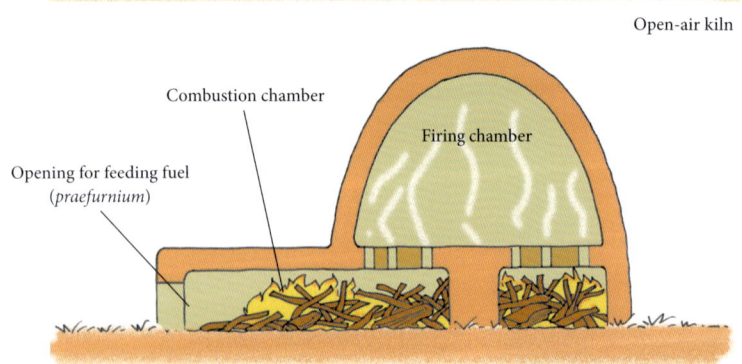

Vertical kiln

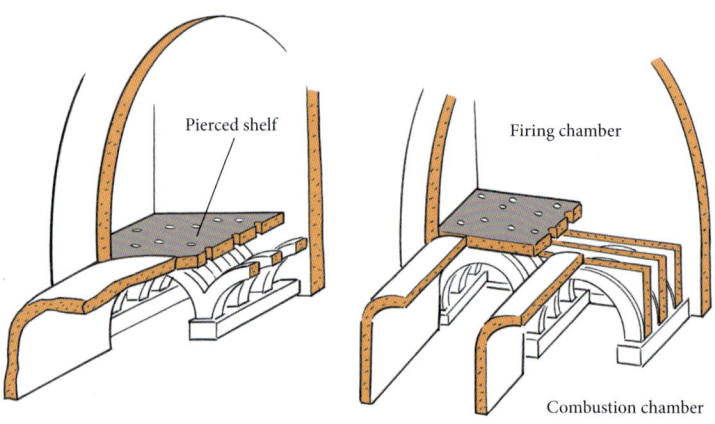

use of baked brick in architecture. However, it was only beginning in the Augustan period that Rome's great brick makers began furnishing material on a large scale. The first monumental complex in the city to be built entirely of baked bricks was the Castra Praetoria, the barracks for the Praetorian Guard built by the emperor Tiberius between A.D. 21 and 23. The use of baked bricks subsequently entered into the architecture of the great imperial residences, beginning with that of Tiberius himself, followed by the Domus Aurea of Nero and the palace erected by the Flavians on the Palatine. From the middle of the third century to the beginning of the fourth, the entire economic system of the empire, including the baked-brick industry, experienced a crisis, leading to the reuse of bricks from earlier buildings, often as horizontal courses alternating with others in stone.

Some phases in the production of earthenware construction material, particularly bricks and tiles, were similar to those related to making mud bricks. Once the mixture of argillaceous soil and mineral and plant aggregates had been prepared, it was given the desired shape through the use of a mold with a wooden frame. The pieces thus produced were then set out to dry, losing part of the water of the clay mixture and acquiring a certain solidity. During the drying stage bricks required frequent turning to keep them from curling as a result of shrinkage. The last stage was firing, which took place in special kilns at a temperature of about 800° C. The addition of straw to the mud mixture ensured that the dilations and contractions that took place during the firing process would not cause cracks or breakage. As the straw burned away it formed minuscule cavities that gave porosity to the interior structure, with the advantage of lightening the compactness of the brick.

The type of kiln (Gr. *kaminos*, Lat. *fornax*) used for baking bricks was similar in all ways to those used for the production of clay tiles and amphorae. Pliny the Elder (*N.H.* 7.194)

The various ways of dividing baked bricks.

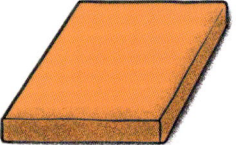
Whole brick

Division along axis

Division along diagonal

Division in four pieces along diagonals

Division in eight pieces along axes and diagonals

Division in sixteen pieces along main and secondary axes and diagonals

attributed the invention of kilns to two mythical Athenian brothers: "Brick-kilns and houses were first introduced by the brothers Eurylas and Hyperbius in Athens; previously caves had served for dwellings." The simplest kind of kiln was the so-called open hearth, set up in a ditch in the open air. Products to be fired were piled in the hearth, on top of and surrounded by a layer of combustible material; this mass was then covered with lumps of sod, clay, and dung, leaving several holes for a draft. Thus the baking was done in direct contact with the fire, at relatively low but not uniform temperatures that resulted in products of low quality.

More evolved were kilns that separated the site for the heat source from the area of the load to be baked. In general, kilns of this type were circular or rectangular in shape; naturally combustible material was inserted through an opening (Lat. *praefurnium*) that led to a combustion chamber, where the material to be burned was piled. Above this was the charge, or firing chamber, where objects for firing were arranged. A flat shelf, with holes to permit the passage of heat, separated the two areas and offered a surface on which to place the pieces to be baked.

The load of material to be baked in the kiln was set upon bricks that were stacked end on end in the firing chamber. As a result, part of the load was located in a lower position, and thus closer to the combustion chamber, while the rest of the pieces were above, in an area where the temperature was less than optimal for firing. The quality of individual bricks consequently varied according to their position in the kiln. Those placed lowest in the load, immediately above the pierced shelf, and to which recent tradition has given the name ironware, were extremely hard because fired at a high temperature, but they were not useful for building masonry walls. Placement above the ironware resulted in "strong" bricks, which were of better quality and strength and suitable for any type of use. Next were "sweet" bricks, of finer quality; and finally, the poorest examples, which were very pale in color because hardly baked and consequently less resistant to static stress.

LARGE-SCALE PRODUCTION: KILNS AND KILN WORKERS

In Rome and throughout the Roman empire, the spread of construction material made of terracotta was favored by several factors, including the speed of the production process, the comparatively low level of specialization required of workers, the abundant availability of clay, and above all the possibility of manufacturing products in a standardized, unvarying shape, easy to transport and put to use.

During the early period of terracotta construction, baked bricks produced in the Greek world roughly duplicated the shapes and sizes of mud bricks. The measurement system adopted in Rome and the Roman world, based on a foot of 29.6 centimeters, led to the production of bricks with a square base in several standardized sizes, known by their Latin names: the *bessalis*, with sides measuring 19.7 centimeters, equal to two-thirds of a Roman foot; the *pedalis*, 29.6 centimeters, equal to one Roman foot; the *sesquipedalis*, 44.4 centimeters, equal to 1.5 Roman feet; and the *bipedalis*, 59.2 centimeters, equal to two Roman feet.

As will be discussed subsequently in chapter 5, the Roman builders generally used bricks to create the two external faces of an individual structure, and then filled the space between the walls with *opus caementicium*, a collection of mortar and various bits and pieces of other material. Bricks were rarely used whole but instead were broken into smaller elements (Lat. *semilateres*) of triangular or rectangular shape, by means of a cut along the axes and/or the diagonals. This technique made possible significant economy at a work site, since a single *sesquipedalis* could be broken up to create from eight to sixteen bricks, each with a triangular base. Laid in courses with the hypotenuse facing outward, the *sesquipedalis* elements appeared on the face of the wall as bricks with a length of 31.4 centimeters (using eight bricks) or 15.7 centimeters (sixteen bricks). Similarly, a *bipedalis* could be divided into eighteen triangular bricks, each with a hypotenuse of 27.8 centimeters, while from a *bessalis* and a *pedalis* were obtained four triangular bricks, each with a hypotenuse equal to one side of the undivided original brick. A single divided brick, installed with the triangular termination hidden inside a wall, thus appeared on the facade as four bricks. The result for the builder was a fine savings!

Two channels with terracotta pipes at the Sanctuary of Heracles on Kos.

The use of bricks was not restricted to the construction of walls. In fact, distinctive brick shapes were made for specific purposes. Bricks in radial form were produced to make columns; circular bricks for use in the hypocausts of bathing establishments; brick with rectangular or polygonal bases (square, rhomboid, or hexagonal) for floorings; and bricks with a trapezoidal section for the curving lintels of arches.

As early as the Hellenistic constructions in Greece and Magna Graecia, the custom spread of impressing bricks with stamps during the drying stage that preceded baking. These stamps were applied with seals (Lat. *signacula*) of wood, terracotta, or metal (iron or bronze); metal seals were usually given a ring or a dado on the back to grip while pressing.

OTHER CONSTRUCTION MATERIALS IN TERRACOTTA

Baked bricks were not the only architectural elements made in the shops that produced bricks used for building. A variety of other types of products were also made of baked clay, and some of them had been in use for several centuries before the spread of bricks;

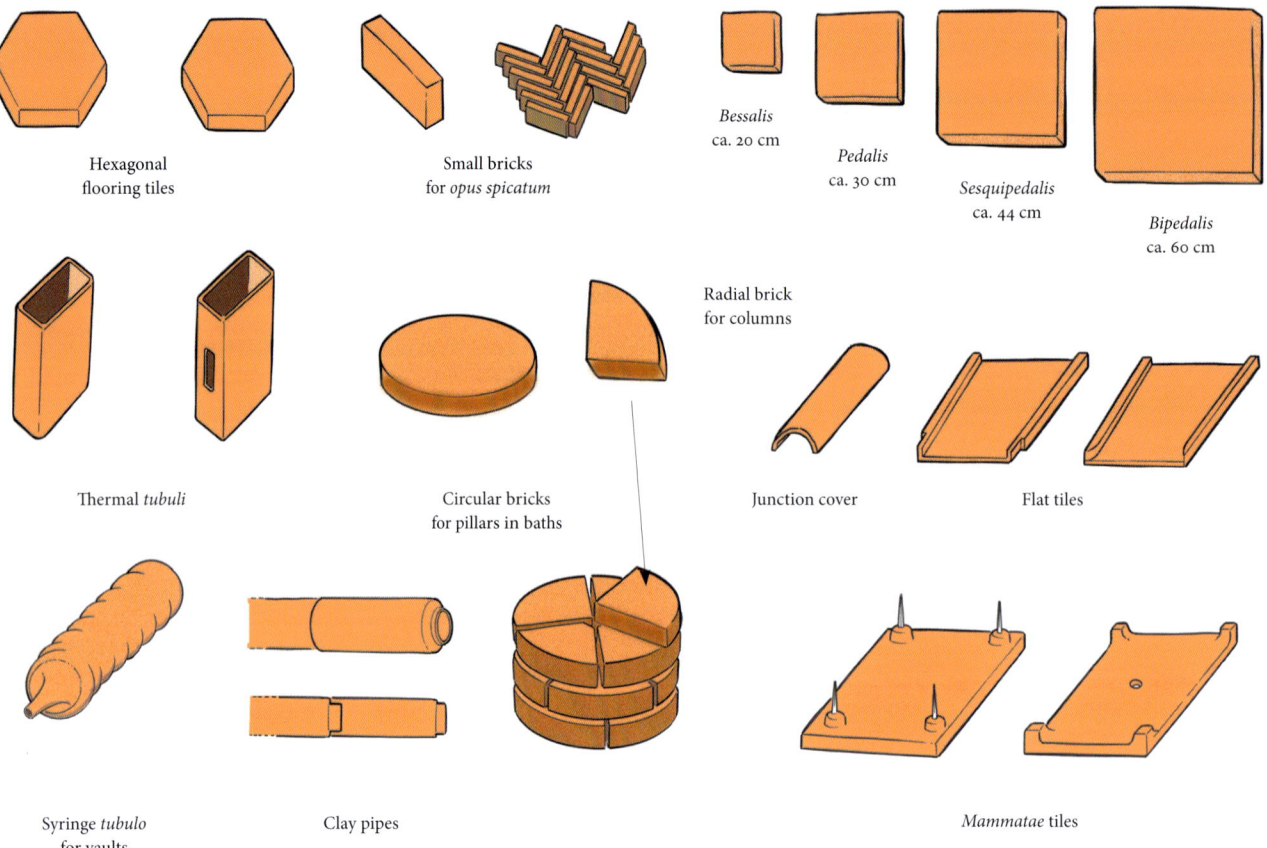

Various clay products made for use in construction.

of particular interest are flat flanged tiles (Gr. *keramides*, Lat. *tegulae*) as well as cover tiles (Gr. *kalypteres*, Lat. *imbrices*). Already present in the Minoan and Mycenaean worlds, these tiles and covers once again appeared in buildings over the course of the great architectural transformation that took place throughout the Mediterranean during the seventh century B.C. We will return to the shape and qualities of these materials in later chapters, within a discussion of the structure of the roofs of ancient buildings.

Yet another group of clay products, equally ancient, was made for a very different use. These were the clay pipes (Gr. *auloi*, Lat. *fistulae*) produced for the construction of water canals and vertical rain spouts. They were made in Crete as early as the Minoan period, and were used in the expansive palace at Knossos. During the course of the Greek Archaic and Classical periods, clay pipes were distinguished by the special care given them throughout the entire production process; although made to be used underground, they were sometimes painted. Fabrication of clay pipes during the Roman period was more standardized, both in terms of the basic mixtures employed and the specific fittings used to connect pipes for channels.

With the Roman period and the development of new construction techniques, other hollow clay products came into production, made to serve additional needs. Among these new forms were small parallelepiped tubes used for heating warm areas in bathing complexes; by creating vertical canals along the walls, these tubes allowed for the passage of hot air. Tubes were also inserted in the *opus caementicium* of vaults to reduce weight or to create a truly self-supporting centering.

LIME, MORTAR, AND PLASTER

LIME

The principal mortars of antiquity employed two types of binder to join blocks and fill gaps between them: lime (Gr. *titanos* or *konia*, Lat. *calx*) and gypsum (Gr. *gypsos*, Lat. *gypsum*).

Lime is produced by heating limestone rocks to a temperature of approximately 900° C. It is not known when humans first learned of the effects of high heat on this type of rock, nor when they began to intentionally heat rocks of this type, but the recognition of lime's value probably resulted from the investigation of an accidental discovery, which was then probably followed by experiments, and trial and error, in a manner roughly similar to the discovery and investigation of terracotta. The process by which lime can be produced was established early in the historical record, and documented in the fourth century B.C. by the Greek philosopher Theophrastus (*On Stones* 2.9): "Some of them [stones] melt and become fluid when subjected to fire.... Some go so far as to say that all of them melt except marble and that this burns up and lime is formed from it."

Marbles and limestones are composed essentially of calcium carbonate ($CaCO_3$) and/or magnesium, and during heating the rocks undergo a calcination process that transforms the carbonate into calcium oxide and/or magnesium and also frees a quantity of carbon dioxide equal to about 44 percent of the total. The calcium oxide thus obtained

Preceding page. Mausoleum of Cecilia Metella (first century B.C.) on the Via Appia in Rome. Note the core of the high base, created with lime mortar and irregularly shaped stones.

Right. The monumental structures of the Domus Augustana, the emperor's residence on the Palatine. They testify to the expertise acquired by Roman builders in the use of lime mortar, both for brick facing walls and for the core in opus caementicium.

is called quicklime (Gr. *konia asbestos* or *titanos akatasbestos*, Lat. *calx viva* or *glaeba*). After heating, the rocks have a powdery appearance and have lost from one-fifth to one-tenth of their original weight. For this reason the calcination operation is most often performed directly at the quarry site, so that the material can then be transported to the work site at a diminished weight. Vitruvius (2.5.3) emphasized the reduction in weight that resulted from heating lime: "Whatever weight the stone possesses when it is thrown into the kiln, it cannot answer to that when it is taken out; but when it is weighed, the bulk remaining the same, it is found to lose about one-third of its weight when the moisture is burnt out."

When used as a binder to make mortar, quicklime is subjected to a process known as slaking, accomplished by immersing the fragments of cooked rocks in tanks full of water. This method leads to hydration of the quicklime, which through high heat (up to 3000° C) is transformed into a pasty substance referred to as slaked lime (Gr. *titanos esbesmene*, Lat. *calx extincta*).

Various ancient sources cite the singularity of the thermal phenomena related to the process of slaking quicklime. Pliny the Elder (*N.H.* 36.174), for example, notes that "lime possesses one remarkable quality: once it has been burnt, its heat is increased by water." More explicit is the passage written some centuries earlier by Theophrastus (*On Stones* 9.66) about the "gypsum" produced by baking limestone in furnaces: "After it has been pulverized and water has been poured on it, it is stirred with wooden sticks; for this cannot be done by hand because of the heat." The reference here is clearly to lime, since gypsum is not subject to slaking. The same author, referring to the production of textiles, adds (9.68), "Its nature is such that [slaked lime] seems . . . to contain the qualities both of lime and of earth, namely, heat and stickiness, or rather each of these in a marked degree. It is also clear from the following example that

it has a fiery nature; for once a ship loaded with clothes was itself burnt when the clothes became wet and caught fire." Apparently, water had come in contact with lime, which was typically used as a mordant to fix the colors of fabrics, and thus set off the process of slaking, resulting in the emission of heat; the fire that ensued could not be extinguished, since the water necessary to put it out had the very opposite effect.

GYPSUM

Gypsum is composed of calcium sulfate with two molecules of water ($CaSO_4 \cdot 2H_2O$). During cooking, gypsum loses part of the water of crystallization, becoming hemihydrate sulfate of calcium ($CaSO_4 \cdot \frac{1}{2}H_2O$). The baking temperatures for gypsum are far lower than those required for lime, and the formation of the hemihydrate is reached at the relatively low temperature of 128° C. Powder is obtained by grinding the baked stone, and when this is mixed with an appropriate quantity of water, the result is a readily malleable plastic mass. Unlike lime-based mixtures, the gypsum mass has a very low coefficient of shrinkage, and it thus can be used without degreasing aggregates. The addition of a small amount of salt or other organic substance has the effect of reducing the time required for setting.

Gypsum was widely used in ancient architecture for the creation of decorative elements like stuccoes and moldings. Pliny the Elder (*N.H.* 36.182–83) wrote of the "affinity between lime and gypsum, a substance of which there are several varieties": "It can be produced from a heated mineral, as in Syria and Thurii; it can be dug from the earth, as in Cyprus and Perrhaebia. There is also that of Tymphaea, which is stripped from the earth's surface. The mineral that is heated ought to be like onyx marble or crystalline limestone. In Syria the hardest stones possible are selected for the purpose and are heated along with cow dung so that the burning may be accelerated."

According to Pliny, the finest gypsum "is prepared from specular stone or from stone that flakes in the same way." Once moistened, he adds, gypsum "should be used instantly, since it coheres with great rapidity," although it can be returned to its previous state: "There is nothing to prevent it from being pounded and reduced again to a fine powder." Pliny notes several applications for gypsum, which can perform as "a serviceable whitewash" and also be "used with pleasing effect for making molded figures and festoons in architecture."

BAKING STONES

The methods used by the ancients for the baking and calcination of stone were carried on in the artisan tradition until recent times, and in fact continue to be used in several less-industrialized countries.

The most ancient method, still employed in some regions of the Near East and Africa, is that of baking the stone in a covered area. A layer of limestone is arranged directly on the ground, and then covered by a thick blanket of animal dung that serves as the combustible. Theophrastus (*On Stones* 9.69) wrote of the methods used to bake marble as well as "the more ordinary kinds of stones": "Cow manure is placed alongside the hardest ones to make them burn better and more quickly. It seems to become extremely hot when it has been set on fire, and stays hot for a very long time." This method, however, was somewhat slow, and called for several days of baking. Despite the use of dung, the temperatures reached were never very high.

More efficient, but also more laborious, was the use of a true furnace (Gr. *kaminos*, Lat. *fornax*), usually shaped like a cone with a circular base from 2 to 7 meters in diameter. The furnace worker (Gr. *chalikokaystes*, Lat. *coctor calcis*) first prepared an area for its installation, most often located at the bottom of a slope, and then set up the base of the furnace, creating a small circular wall of large

Schematic reconstruction of the lime kiln described by Cato.

limestone rocks topped with a round dome. As in ceramic kilns, this combustion chamber served to contain the fire for baking. Having completed this phase, the artisan moved on to loading the furnace by arranging the stones in the shape of a cone, beginning with the largest fragments and ending with the smallest, which required less baking and could be located at the top of the pile. The combustibles to be used required the capacity to release strong heat with high flames, so preference was for small, dry materials.

The construction and use of ancient lime furnaces was described in detail during the first half of the second century B.C. by the Roman statesman Cato the Elder (*On Agriculture* 38):

Build the lime kiln ten feet across, twenty feet from top to bottom, sloping the sides to a width of three feet at the top. If you burn with only one door, make a pit inside large enough to hold the ashes, so that it will not be necessary to clear them out. Be careful in the construction of the kiln; see that the grate covers the entire bottom of the kiln. If you burn with two doors there will be no need of a pit; when it becomes necessary to take out the ashes, clear through one door while the fire is in the other. Be careful to keep the fire burning constantly, and do not let it die down at night or at any other time. Charge the kiln only with good stone, as white and uniform as possible. In building the kiln, let the throat run straight down. When you have dug deep enough, make a bed for the kiln so as to give it the greatest possible depth and the least exposure to the wind. If you lack a spot for building a kiln of sufficient depth, run up the top with brick, or face the top on the outside with fieldstone set in mortar. When it is fired, if the flame comes out at any point but the circular top, stop the orifice with mortar. Keep the wind, and especially the south wind, from reaching the door. The calcining of the stones at the top will show that the whole has calcined; also, the calcined stones at the bottom will settle, and the flame will be less smoky when it comes out.

After baking, the lime would be one of two types, depending on the quality of limestone that had been used: "fat lime" (Gr. *asbestos pachys* or *asbestos aspros*, Lat. *calx pinguis*) was derived from baking pure limestone, while "lean lime" (Gr. *asbestos ischnos*, Lat. *calx macra*) was obtained from limestone that contained 1 to 6 percent argillaceous impurities. The Romans were highly skilled at the production of lime and the evaluation of its qualities, as indicated by the prescriptions contained in ancient sources. For example, Vitruvius (2.5.1) noted that the qualities of each particular type of lime suited it to specific production methods: "Lime which shall be out of thick and harder stone will be useful in the main structure; that which shall be of porous material, in plaster work." Several decades later Pliny the Elder (*N.H.* 36.174) also noted the applications of different qualities of lime: "As for lime, Cato the Censor [the Elder] disapproves of preparing it from variegated limestone, for white limestone produces a better quality. . . . It is more serviceable if it is produced from quarried stone than from stones collected on the banks of rivers. A superior kind is made from stone used for querns [millstones], for they have a certain unctuous character."

MORTAR AND ITS INGREDIENTS

When lime receives water and sand and then strengthens the structure, the following seems to be the cause: just as other bodies, so also stones are blended of the elements. And those which have more air are soft; more water, are pliant from the moisture; more earth, are hard; more fire, are more fragile. Therefore if stones of this last quality are crushed before they are burnt, and mixed with sand, and thrown into the work, they do not become solid, nor can they hold the building together. But when they are thrown into the kiln, they are seized by the violent heat of the fire and lose the virtue of their former solidity. Their strength is burnt out and exhausted and they are left with open and empty pores. Therefore when the moisture which is in the body of that stone, and the air, are burnt out and removed, and the stone retains the remaining latent heat, on being plunged into water . . . the moisture penetrates into the open pores, and it seethes and thus, being cooled again, it rejects the heat from the substance of the lime. . . . When the pores and attenuations of the lime are open, it catches up into itself the mixture of the sand; thus it coheres and, as it dries, joins with the rubble and produces solid walling.

With these words Vitruvius (2.5.2–3) sought to clarify the effect of baking on limestone's chemical and physical properties. The lime thus produced in fact possessed the capacity to act as a binder in various types of mixtures. Until a few centuries ago it was the most suitable binder for architecture.

The base for lime used in building is composed of slaked lime (Lat. *calx extincta*), a dense paste created by mixing lime with a quantity of water less than 20 percent of the total. Further dilution with water results first in "lime milk" (20 to 30 percent water) and then "lime water" (as much as 70 to 80 percent water). While there are occasional ancient references to the use of lime milk (Gr. *gypsodes gala*, Lat. *calx diluta*) for the creation of brushed plastering on stone or terracotta, the architectural use of slaked lime seems to have been limited to the creation of thin layers for the positioning and horizontal arrangement of squared blocks. This technique appears in various Roman constructions dating to both the republican and imperial periods. In Rome, for example, slaked lime was used in building the Pons Aemilius, the Tabularium, and the Temple of Veiovis.

The primary use of lime in construction was to make mortar (Gr. *koniama*, Lat. *materia*), a mixture formed by the addition of water and medium-fine aggregates, usually composed of sand (Gr. *ammos*, Lat. *harena*) with a particle-size measurement of less than 2 millimeters. The qualities of the mixture depended not only on those of the lime used but also on the nature of the inert additives. Vitruvius (2.4.1–3) pays close attention to the types of sand most appropriate for mortar, noting that one "must first inquire about the sand, that it be suitable for mixing material into mortar, and without the admixture of earth." He notes that there are four kinds of quarried sand: black, white, red, and that derived from lignite. In all cases, writes Vitruvius, "that which makes a noise when rubbed in the hand will be best; but that which is earthy will not have a like roughness. Also, if it is covered up in a white cloth, and afterwards shaken up or beaten, and does not foul it, and the earth does not settle therein, it will be suitable." When sand cannot be obtained from a pit, "it must be sifted out from the riverbed or from gravel . . . [or] from the seashore." Sand from these sources, however, has these defects: "It dries with difficulty, nor does the wall allow itself to be loaded continuously without interruptions for rest, nor does it allow of vaulting. But in the case of sea sand, when plastered surfaces are laid upon walls, the walls discharge the salt of the sands and are broken up." In contrast, sand from a quarry "quickly dries in buildings, and the surface lasts; and it admits of vaulting, but only that which is fresh from the pit. For if after being taken out it lies too long, it is weathered by the sun and the moon and the hoar frost, and is dissolved and becomes earthy. Thus when it is thrown into the rubble, it cannot bind together the rough stones, but these collapse and the loads give way which the walls cannot maintain." However, Vitruvius notes that while quarried sand "has such virtues in buildings, it is not useful in plaster

Detail of the facade of the Tabularium, the large structure built in Rome during the first half of the first century B.C. to hold the state's public archives. The opus quadratum *is characterized by thin layers of lime between the blocks.*

work; because owing to its richness, the lime when mingled with straw cannot, because of its strength, dry without cracks."

Less than a century later Pliny the Elder (*N.H.* 36.175) limited himself to stating simply: "Of sand, there are three varieties: there is quarry sand, . . . river [sand], or alternatively sea sand."

In fact, ancient builders—in particular those of the Roman period—experimented widely with another type of aggregate for the production of certain special mortars, including the so-called pozzolana (Lat. *pulvis puteolanus*). The addition of these volcanic lapilli, found in Italy both in the area of the Phlegraean Fields (in a pale gray color) and in the Roman countryside (in colors from black to violet-red), gave mortars excellent hydraulic qualities. Vitruvius (2.6.1) wrote of "a kind of powder which, by nature, produces wonderful results. It is found in the neighborhood of Baiae and in the lands of the municipalities around Mount Vesuvius. This being mixed with lime and rubble, not only furnishes strength to other buildings, but also, when piers are built in the sea, they set under water." In addition, he presented a theory regarding the origins of pozzolana: "Now this seems to happen for this reason: that under these mountainous regions there are both hot earth and many springs. And these would not be unless deep down they had huge blazing fires of sulfur, alum, or pitch. Therefore the fire and vapor of flame within, flowing through the cracks, makes that earth light. And the tufa which is found to come up there is free from moisture. Therefore, when three substances formed in like manner by the violence of fire come into one mixture, they suddenly take up water and cohere together. They are quickly hardened by the moisture and made solid, and can be dissolved neither by the waves nor the power of water."

The use of pozzolanic mortar was widespread in Roman architecture, and adopted for a broad range of uses, not just for those

Left. Trajan's Market, erected early in the second century A.D. immediately northeast of Trajan's Forum, was built with an adhesive lime mortar mixed with high-quality gray pozzolana.

LIME, MORTAR, AND PLASTER 69

Top. Blocks of gypsum piled in the House of the Iliac Chapel at Pompeii. This material was to have been used for the restoration of structures damaged in the earthquake of A.D. 62.

Bottom. Preparation of lime mortar.

instances when its hydraulic capacities were required. Mixtures were made using high-quality raw materials that were aged and blended following precise formulas.

PREPARATION OF MORTAR

In ancient times, just as in the historical era, the preparation of mortar for use in masonry constituted a crucial step toward the successful outcome of a construction project. An error in the quantity of any of the base components (lime, sand, or water) could result in the collapse of an entire structure. The use of too much sand, for example, resulted in mortar that was too thin, while an insufficient quantity of sand resulted in mortar that was too coarse. Vitruvius (2.5.1) provided a formula that was subsequently repeated by Pliny the Elder (*N.H.* 36.175): "When it [the lime] is slaked, then let it be mingled with the sand in such a way that if it is pit sand, three [parts] of sand and one of lime is poured in; but if the sand is from the river or sea, two of sand and one of lime is thrown together. For in this way there will be the right proportion of the mixture and blending." These instructions reflect the ancients' clear understanding that the addition of various kinds of aggregates, including ground pottery as well as bricks, made it possible to render mortar even more durable. In this regard Vitruvius further noted that "if anyone adds crushed and sifted potsherds in the proportion of one to three, he will produce a blending of material which is better."

Mortars were usually prepared at the construction site. Having located a large enough space, the mason piled up sand—if necessary, first passing it through a sieve (Gr. *koskinon*, Lat. *cribrum*)—and worked the mound into the shape of a crater; he then poured slaked lime into the center and added small amounts of water, until the correct proportion was achieved. The ingredients were mixed with the use of a tool (Gr. *makella*, Lat. *ligo*) similar to a hoe with a very long handle, to eliminate lumps and to work the sand and lime into a homogeneous mixture.

The eruption of Mount Vesuvius in A.D. 79 buried the nearby cities and brought a sudden end to the various work sites that had been established for the restoration of buildings damaged more than a decade earlier by a powerful earthquake. Excavations and research have made it possible to collect a great deal of information on the technical and organizational aspects of these sites. For example, the

The large public fountain on the western corner of the Baths of Neptune at Ostia. The structure still preserves a good part of its ancient covering in mortar mixed with ground-up earthenware.

lime that was to be used for restoration work at the House of the Moralist at Pompeii was piled in the entry hall, in a site sheltered from rain. At other work sites in the city, such as those set up at the House of the Iliac Chapel and at the structure designated as Home V.3.4, excavators found amphorae filled with lime, from which the upper portions had been broken away, suggesting that it was customary to carry slaked lime for mortar inside such containers. Other traces of work sites with piles of slaked lime have been identified at Hadrian's Villa at Tivoli and at the Villa dei Quintili on the Via Appia, where an entire room full of this material was found. Of course, as lime was the most costly component of mortar, it was not uncommon for building contractors to seek savings in its use. Pliny the Elder (*N.H.* 36.176) offered his opinion on the consequences of such practices: "The chief reason for the collapse of buildings in Rome is the purloining of lime, as the result of which the rough stones are laid on each other without any proper mortar."

OPUS SIGNINUM MORTAR

As we have discussed, one particular type of mortar called for adding to the principal components (lime, sand, and possibly pozzolana) an aggregate of an earthenware nature derived

LIME, MORTAR, AND PLASTER 71

Remains of the ancient city of Iaitas are preserved on Mount Iato in western Sicily. The so-called Peristyle House I had a bathroom with a masonry tub. All the walls of the space were covered with mortars mixed with ground earthenware of different particle sizes. The floor was made of the same material and decorated by the insertion of small white-marble tesserae.

from breaking up construction materials. This mortar, called *opus signinum* in Latin, had certain hydraulic properties and was used in all situations in which it was necessary to waterproof a structure. It was therefore employed to cover water cisterns and tanks, ordinary floors, and even more often the suspended floors (Lat. *suspensurae*) used in the heated areas of bathing complexes. Vitruvius (7.1.3) prescribes covering the middle layer (Lat. *nucleus*) of the floors with "a hard coat of powdered pottery . . . three parts to one of lime."

Mortar made with powdered pottery has very ancient origins. As early as the tenth century B.C., a mixture of this type was used to waterproof several cisterns in Jerusalem, and the technique was probably introduced to the West by Phoenician builders. It appears that the Greeks had begun exploiting the capacities of mortar mixed with ground pottery at least as early as the fourth century B.C., as indicated by the bath installation in the Sanctuary of Asclepius at Gortys in Arcadia and by various homes at Olynthus. During the Hellenistic period the technique spread across wide areas of the Mediterranean, from the Greek world (Delos and Thera) to the Punic (Tharros, Selinunte in the Carthaginian period, and Solunto). In Pompeii it was present as early as the Samnite period, and in Rome from at least the third century B.C. Mortar with ground pottery was also used for the application of plaster, especially inside rooms exposed to humid conditions. Vitruvius (7.4.1) is the principal source for information on this technique: "I will now explain how stucco is executed in damp places so as to avoid blemishes. And first as to rooms on the level ground. To the height of about three feet from the pavement, rough-cast made of powdered earthenware instead of sand, is to be laid on, so that this part of the plaster may not suffer from damp." Archaeological excavations have indicated that the instructions given by Vitruvius were widely followed in Roman architecture, as attested by numerous examples in Rome and Vesuvian cities, as well as in Gaul and the eastern provinces.

MORTARS FOR PLASTER AND STUCCO

The primary use of mortar was related to the building of masonry structures, but other mixtures were specially produced for covering walls, both to decorate surfaces and to protect them from atmospheric agents, such as rain and wind.

In the Greek world the custom of covering plaster (Gr. *dorosis*, Lat. *opus tectorium*) was rooted in the Minoan and Mycenaean cultures. From the Archaic period plaster for both sculpture and architecture was covered with stucco (Gr. *koniama*, Lat. opus *albarium*). Distinguishing between the two techniques is not always easy, nor is the terminology always clear. In general, plaster refers to the use of mortar with a base of lime and sand, while stucco calls for a mixture of lime, calcined gypsum, and marble dust.

The application of a layer of plaster to dress architectural surfaces appears in certain large temple creations as early as the seventh century B.C., but the technique only spread widely during the following century. Initially, plaster was used to cover walls, a system that reflected the tradition of protecting walls made of perishable materials, for example, in

the Temple of Apollo erected at Corinth in the seventh century B.C. As discussed, the process of building with stone developed through the use of locally available materials, in particular tufa and limestone. Consequently, it soon became common practice to cover architectural elements with a mixture based on lime and marble dust (or limestone dust), Such a strategy allowed builders to achieve a level of detail similar to that offered by marble. The Sanctuary of Athena Pronaia at Delphi—the oldest temple in tufa, erected in the first decade of the sixth century B.C.—as well as

Top. Outer wall of the House of Euthycus at Pompeii (Home VI.11.8). The plaster covering, laid on in several layers, was decorated to simulate a wall in opus isodomus.

Bottom. Tombs in the necropolis of Porta Nuceria at Pompeii. They were originally decorated with relief plasters that imitated a wall made with squared blocks.

Top. Preparation of a wall decorated with frescoes. Note the presence of several layers of plaster.

Page 76, top. Roman masons produced mortars of high quality and great strength. Their endurance was such that they still maintain imprints of the slabs that were set in them, even preserving negative impressions of letters from recycled slabs that bore inscriptions, as in this section of flooring from the baths at Copia in Calabria.

Page 76, bottom. The precious stuccoes of the theater at Ostia. Created in the second half of the second century A.D., they were decorated with deeply incised rosettes and figures of flying Horae.

the building that replaced it around the end of the century were distinguished by columns that featured shafts and capitals covered with a thin layer of fine beige stucco. A similar technique was employed at many other work sites throughout the Greek world (for example, at Aegina, Athens, Corinth, and Olympia) and Magna Graecia (at Poseidonia [Roman Paestum] and Metapontum) between the sixth and the fifth centuries, as well as at a number of temple structures in Sicily (at Agrigento, Selinute, and Syracuse). In fact, beginning in the fourth century B.C. most public monuments and private structures not made in marble received a decorative dressing.

The ancient technique employed to make this decorative plaster dressing did not differ greatly from the methods used by today's masons. Mortar was mixed in a wooden trough (Gr. *skaphē*, Lat. *mortarium*) in quantities that naturally were far more restrained than those prepared for masonry constructions. The artisan (Gr. *koniates*, Lat. *tector*) first prepared the mixture and then used a trowel to place a small amount (Gr. *hypagogeus*, Lat. *rutrum*) on a wooden tablet, known as a float, with a handle on the back. After spreading a layer of dressing, the surface was finally regularized with a tool known as a smoother (Gr. *tripter*, Lat. *liaculum*).

In most cases the creation of wall dressings involved the application of several superimposed layers, differentiated by the material used and the particle size of the aggregates. The first stage was roughing out (Lat. *trullissatio*), to remove any unevenness on the wall and at the same time abrade it so that the plaster could be better anchored to the wall surface. Mortar for this layer was somewhat coarse, often made with unsieved sandy aggregates and hay or ground earthenware. To this roughed-out surface the artisan then applied a first coat (Lat. *harenatum*), composed of one or more layers of mortar that had been produced with finer sand, and in many cases with the addition of pozzolana. Next came the finishing layer (Lat. *politiones*), made with lime mortar and marble dust. While this last and finest layer of plaster was still damp, the application of color (Gr. *chromata*, Lat. *colores*) began. Painting of this type, called fresco ("fresh") because it was performed on a layer that was still damp, permitted the pigments to penetrate the mortar and become sealed beneath the surface film, remaining vivid and unchanged over time.

The technique of plastering outlined above, and carefully described by Vitruvius (7.2–3), was adopted by builders of both Greek and Roman architecture. Many of the Hellenistic homes on Delos have walls covered by plaster that is composed of two, three, or even four superimposed layers, and the adherence between them was improved by means of parallel or patterned incisions made with a trowel. The thickness of each level of plaster varied to correspond with the quality of the mortar used. On average, the lower layer maintained a depth of between 3 and 15 centimeters; the intermediate layers, between 1 and 1.5 centimeters; and the final layer, between 0.3 and 0.5 centimeter. Similar measurements

View of the peristyle in the House of the Trident on Delos. The building preserves large portions of its painted plaster in the architectonic style.

were found for plaster layers at Priene, while between the fifth and fourth centuries B.C., the preference in some homes at Olynthus was to apply a single layer of dressing, at a thickness of approximately 0.5 centimeter, directly to the mud-brick walls. It was not uncommon to add an aggregate of volcanic origin to the dressing mortar, thus giving the mixture greater endurance. This method can be seen in the wall covering of the Temple of Athena and Hephaestus in Athens, where pozzolana from the island of Santorini was added to the mixture, along with cut hay.

Procedures for executing relief stuccoes (Gr. *koniatika*) differed from those used for various types of wall dressings, including not only the grooved coverings of columns and the imitation of *opus quadratum* of both the architectonic style and the First Style at Pompeii, but also cornices and decorative figural elements. In these cases the material used for mortar (lime, marble dust, and calcined gypsum) had to be of the very highest quality. Molding was created with the help of templates that reproduced the desired profiles, while reliefs were produced by molds that were pressed into the still damp stucco.

One of the most ancient and important examples of relief stucco is the decoration of the Tomb of the Judgment of Lefkadia. This large chamber tomb, dated around 300 B.C., features a monumental double-order facade that includes a decorative frieze with figures in polychrome stucco. The vault of the *tepidarium* of Pompeii's Forum Baths also preserves a fine example of polychrome stucco, organized with figural lunettes between which are relief depictions of candelabra.

SPECIAL MORTARS

A distinctive group of mortars used in the ancient world included mixtures employed as mastics or as true decorative enamels. The short poem "The Shield of Heracles," at one time attributed to the Greek poet Hesiod, refers to the use of materials of this type in the decoration of the hero's mythical shield (139–43): "In his hands he took his shield, all glittering: no one ever broke it with a blow or crushed it. And a wonder it was to see; for its whole orb shimmered with enamel and white ivory and electrum, and it glowed with shining gold."

Other mortars were made with slaked lime that had been enriched with the addition of organic substances such as oil, casein, or ash. Pliny the Elder (*N.H.* 36.181) offers a recipe for one such mortar: "Maltha is prepared from freshly calcined lime, a lump of which is slaked in wine and then pounded together with pork fat and figs, both of which are softening agents. Maltha is the most adhesive of substances and grows harder than stone. Anything that is treated with it is first thoroughly rubbed with olive oil." These mixtures were produced when extraordinary endurance and resistance were required, as in the case of hydraulic joints. The fourth-century Roman writer Marcus Cetius Faventinus (*Vitruvius and Later Roman Building Manuals* 6), for example, prescribed sealing the junctures of terracotta pipes used in water conduits with a mixture of lime and oil. In his wake the sixteenth-century Venetian architect Andrea Palladio (*Four Books of Architecture* 1.40.q-2) distinguished among the kinds of mortars to use in different rooms of a bathing complex, depending on the temperature of the bath water. To build a room intended for warm water, mortar was produced with equal quantities of hard pitch and white wax mixed with tow, liquid pitch, ground brick, and finely ground lime, or by combining liquefied amoniac gum, figs, tow, and liquid pitch; for cold-water rooms mortars were made by mixing fine lime with ox blood, and adding iron scoria.

CONSTRUCTION TECHNIQUES IN THE GREEK WORLD

WHEN STONE WAS AN EXCEPTION IN ARCHITECTURE: THE SETTLEMENT AT KARPHI IN CRETE

As archaeological research has demonstrated, the end of the Minoan and Mycenaean periods marked the loss of most of the technical knowledge and many of the construction skills that had been applied to the creation of great palaces, and that were associated with the expert and monumental use of stone as a construction material. The buildings of successive centuries, from the twelfth until at least the middle of the eighth century B.C., were made almost exclusively from so-called light materials, those widely available in nature. The reasons for this phenomenon were multiple and remain the subject of study by historians, but one important factor was certainly the social organization of new communities. Each of these small settlements was governed by a village chief (Gr. *basileus*), whose home represented the headquarters of a community's principal religious practice. The great Mycenaean construction sites—intricately structured and teeming with skilled, variously specialized workers who made possible the use of Cyclopean blocks of stone—were by then unimaginable. The construction of a house was generally a family affair, on occasion expanded to assistance from other members of the village. The use of stone, often mere rocks or chunks of irregular shape, was limited to the area of the foundations and the masonry socle that protected the upper structure from dampness. The only other materials used were clay, wood, and additional plant elements, often combined to increase the strength of a structure and to protect walls from atmospheric damage.

In contrast, Crete presented a singular exception, one probably related to an insular conservatism that was maintained into later centuries. Stone never lost its role in that region as the principal building material, perhaps partially in keeping with the ancient Minoan past. Although confined for the duration of the Protogeometric and Geometric periods to modestly sized rectangular buildings (which often served multiple functions), architectural structures on Crete managed to elaborate particularly interesting technical solutions. Against this background the settlement at Karphi assumes a leading role.

The site was discovered by the British archaeologist Arthur Evans in 1896, on mountains around the Lasithi plateau in the central-eastern sector of Crete. The settlement arose, most likely as a place of refuge, around the Vitzelovrysis spring about 1150 B.C., only to be abandoned after about two centuries. Located at an altitude of roughly 1,100 meters above sea level, but in an area sheltered from the cold winds that would have made its occupation otherwise difficult during some months of the year, the site enjoyed the nature of an elevated fortification.

Bottom. The layout of the settement at Karphi is clearly rendered in this general map of the site, made by British archaeologists in the first half of the twentieth century.

Although the British excavations of the first half of the twentieth century uncovered only one-third of the ancient city, archaeologists succeeded in identifying many characteristics of the settlement's organization and the various stages of its construction. They determined that during an initial period inhabitants had limited themselves to erecting simple shelters in rock cavities that occurred naturally or had been dug, but that the settlement had grown with the construction of numerous homes and a network of paved streets that expanded at some points to form small squares.

The architectural aspects of these buildings reveal that the inhabitants of Karphi adopted a variety of technical strategies. Often resting directly on bedrock, homes had no need of foundations. The walls were dry, made without mortar, using pieces of limestone collected in the surrounding area and set in place without any distinction in terms of size. Immense blocks weighing hundreds of kilograms were laid in courses together with very small stones. Unlike buildings in other parts of the Greek world, those at Karphi do not seem to have made use of mud bricks, perhaps in part because of the wide availability of stone. It was probably for this reason that neither the interior nor the exterior walls of rooms were given a protective layer of plaster. A certain discrimination in the selection of stone blocks is present only in the construction of doorjambs and thresholds, the latter often raised above street level to prevent rainwater from entering.

Worthy of note in terms of the building materials used at Karphi is the presence of several squared blocks. These appear to be composed of a kind of stone that is well suited to working—unlike the pieces of stone used for construction of walls—and perhaps they were brought to the village from the nearby site of Karphiotis. Apparently, even in this early phase there was interest in seeking out quarries from which to extract stone materials for use in buildings, despite the difficulties involved in transportation.

Archaeologists have sought to reconstruct the roofing system of these buildings. Roofs were almost certainly flat and supported by horizontal rafters; in some cases, primarily those where the width of the room was too

great to span without internal supports, rafters rested on internal pillars. It is believed that the structure of the typical roof was made from a rough framework with a layer of argillaceous earth that held the slabs of rock that formed a roof covering; an additional, waterproofing layer, about 20 to 25 centimeters thick, served as a cover. It is likely that some kind of hole in the roof allowed smoke to escape from the hearth inside the house. This hole could have been reinforced around its edges by the insertion in the roof of a large jar (Gr. *pithos*) that was transformed into a large chimney by the elimination of its bottom and a portion of its side walls.

Among the constructions at Karphi the temple building located at the northern end of the archaeological area stands out for its architectural significance. Overlooking an open area of roughly 17 by 20 meters, it was initially composed of a single room (4.75 by more than 8 meters); only later, during a second stage of construction, were other rooms added along with a large internal bench on which stood several cult statues. It is interesting that at least part of this construction was undertaken using a building technique that differed from the one used in other constructions at Karphi: the temple walls were made from large blocks of irregularly shaped limestone, with smaller chips inserted to fill in spaces.

ZAGORA: A CITY BUILT OF STONE CHIPS IN THE HEART OF THE CYCLADES

The buildings at Karphi, along with other Cretan constructions from the Protogeometric and Geometric periods, do not constitute isolated episodes, despite the widespread use of mud bricks and other light materials in the Greek world, as discussed in earlier chapters. During the same period other cities also took advantage of an abundance of various kinds of stone, and used that material not only for making foundations but also for building entire structures. Among the best-known examples are those found on the islands of the central-eastern Aegean. The example of Zagora on Andros, located along the maritime routes between Greece and the coasts of the Near East, is distinctive due to its history as well as its finely preserved masonry, allowing us to consider technical aspects of the site's architecture within the broader context of its role as a trading center during the Geometric period.

Inhabited only during the ninth and eighth centuries B.C., the city arose on the island's west coast, on a promontory whose cliffs provided a natural fortification. Thus only along the short northeastern isthmus was there a need to build a stretch of defensive walls, which run for a length of about 140 meters and in some areas are more than 7 meters thick. Within these walls a gate about 4.5 meters wide was opened to permit access by way of a raised street supported on a terraced wall.

The settlement underwent two chronologically close stages of construction and was composed of rectangular houses, often with foundations built directly on exposed bedrock and generally leaning against one another. Inside some of the houses was a rectangular

Bottom. Diagram of the excavations at Zagora. The town developed during the Geometric period, on a large plain with natural defenses on three sides and a stretch of powerful walls to protect the northeast.

Opposite, bottom. Reconstruction of the temple edifice at Karphi, with indication of the roofing system employed.

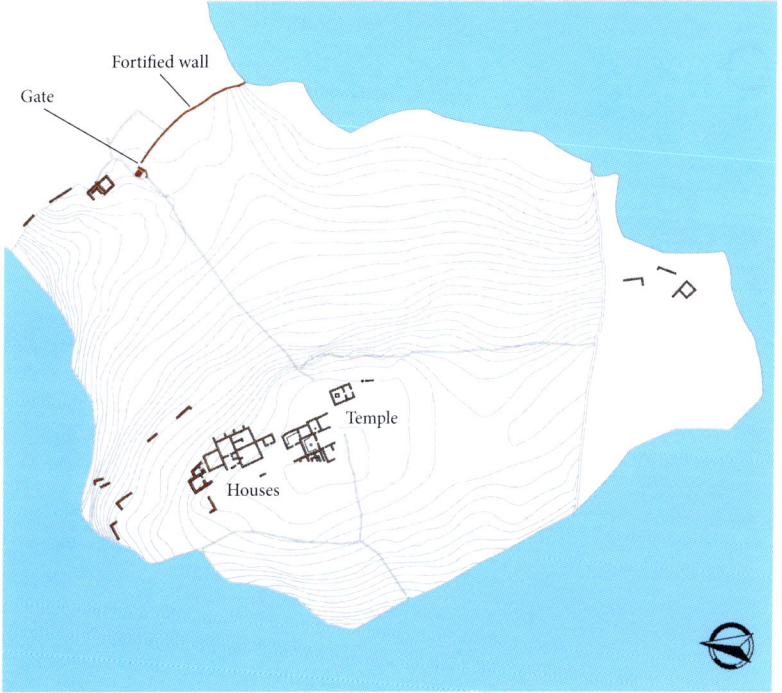

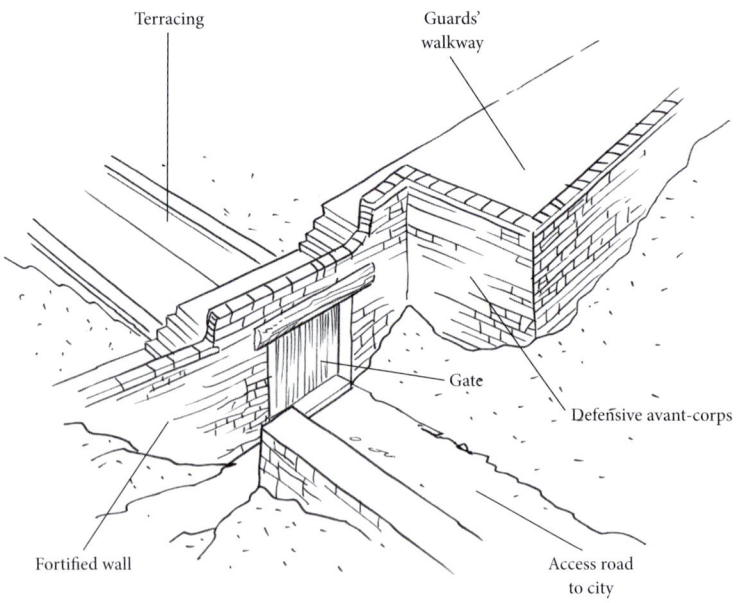

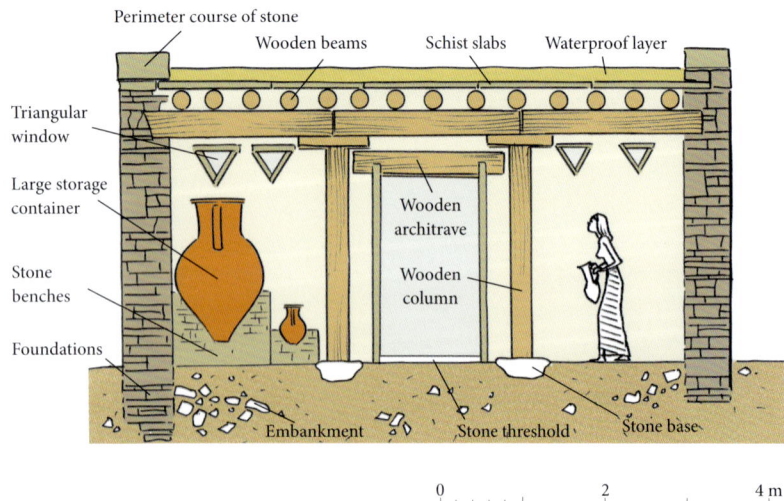

Top. Reconstruction of the fortified wall that gave access to the settlement of Zagora.

hearth made from slabs of schist; many houses also had external open courtyards.

The temple at Zagora, which replaced a primitive sacred enclosure from the Geometric period, stood completely isolated on the central area of the plain. It was of the closed-vestibule type (Gr. *prodomos*) with an almost square cella. Within was the base for an altar or a cult statue. The building, perhaps dedicated to Athena, was made in the first half of the sixth century B.C., and thus dates to a period after abandonment of the settlement.

The masonry of the city's structures was composed entirely of schist, which is widely available on the island; however, in some cases, such as the northern tip of the fortifications and in a few houses, schist was used together with local gray marble, a material that is difficult to work. The building technique employed throughout Zagora involved the creation of two facing walls from relatively smooth pieces of schist that were arranged in roughly horizontal courses, with smaller stones inserted in the interstices; the hollow space between the two walls was then filled with rubble of various shapes and sizes, mixed with mud. To strengthen the connection between the two walls and the rubble core, masons at Zagora inserted, at various points, transverse stones that crossed the full width of each wall. In addition, workers reinforced the corners of walls by piling up larger chunks of stone, which were arranged to face in alternating directions.

At Zagora the wall surfaces, as at Karphi, do not present any covering in clay or plaster, and floors were made of beaten clay, either red or yellow in color. The doors of many constructions were of a roughly uniform width (40 to 50 centimeters). The doorjambs were composed of stone elements inserted directly in the walls, while the architraves were probably wooden. The state of preservation of the walls has made it possible for archaeologists to hypothesize the existence of small triangular windows that served the purposes of ventilation and internal illumination, and were especially useful in the case of houses surrounded on all sides by other constructions.

Roofs were flat, organized by a system of wooden beams and rafters, and supported by internal pillars in instances where rooms were too wide to be spanned easily. Following a method still used in some villages on the island, roofs themselves were made of slabs of schist with a thick covering in clay. It seems that a sort of cornice, also made of pieces of schist, ran along the perimeter of each roof to reinforce it against the strong northern winds that often strike Zagora, in particular on the island's promontory.

MINIATURE ARCHITECTURE: THE CONTRIBUTION OF CLAY VOTIVE MODELS TO OUR KNOWLEDGE OF GREEK BUILDINGS

As a result of certain methods of worship, the ancient world has handed down to us a collection of small terracotta architectural models that are accurate miniature reproductions of actual buildings. The creation of these models was an ancient tradition, with roots in the Minoan and Mycenaean periods. During the Geometric period and then the Archaic, these objects, often richly decorated, were produced to be given as ex-votos in sanctuaries. Today the historical and artistic value of these works is combined with their great documentary significance, for they make it possible to visualize the shape of architectural typologies at the foundation level and to hypothesize about the construction systems behind them.

In the sanctuary of Hera Akraia ("of the Hill") at Perachora, on the promontory that divides the Gulf of Corinth from Livadostro Bay, archaeological excavations in the area of the oldest temple have brought to light several clay models of buildings, all made in Corinth. The most interesting of these, about 35 centimeters long, is dedicated to the goddess and was most likely made around the middle of the eighth century B.C. Set on a rectangular base, it depicts a small apsidal building, preceded by a shallow portico that is supported on the facade by two pairs of twin columns. Above the entrance door, which is set on a slightly raised threshold and tapers upward, the front wall has three small square windows; there was probably a further opening on this wall, between the slopes of the roof. Despite the

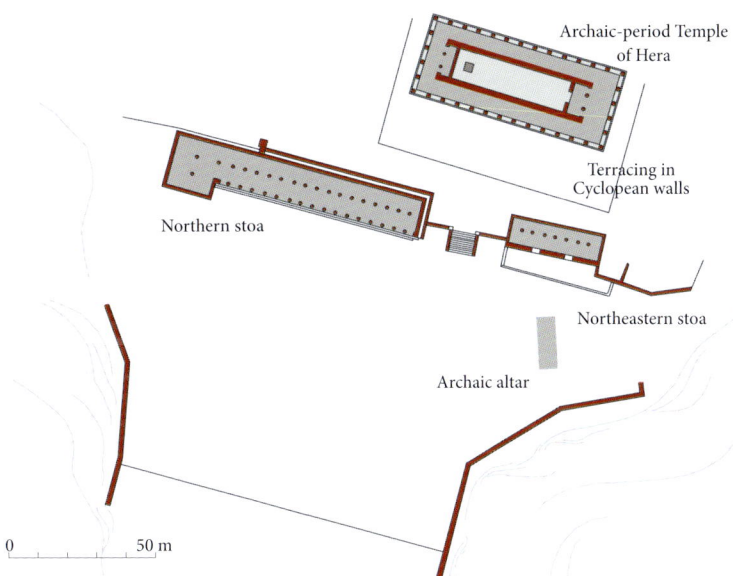

piece's fragmentary state, a painted decoration of geometric motifs can be discerned. There remains some question of whether the clay model is in fact a reproduction of the sanctuary temple, which also has an apsidal shape. In the absence of confirmation, this possibility remains an intriguing hypothesis, especially since the other clay models found in the area also reproduce apsidal buildings.

Another terracotta model, also of considerable size, was found at the suburban Sanctuary of Hera at Argos. Fully 54 centimeters tall, it depicts a building of the megaron type, with a rectangular space and a portico supported by two pillars. The roof had two step ridges and a large rectangular opening in the front gable. Dated to the first quarter of the seventh century B.C., the model bore rich painted decoration, on the slopes of the roof, where the typical step-meander Argive motif appeared, and on the vertical walls, where there are indications of the presence of wooden reinforcing pillars. The band of alternating black-and-white squares that runs along the base of the roof might indicate the placement of roof beams or open spaces between them.

These two models, from two different sanctuaries in Greece, have allowed archaeologists to formulate intriguing theories concerning

Left. Arrangement of the Sanctuary of Hera that lay outside the city of Argos in the Archaic period.

Opposite, bottom. The homes at Zagora show that during the Geometric period advanced technical skills were also applied to domestic architecture. The image presents a cross section of one of the homes, and notes several details of the roofing system.

the construction methods of the buildings depicted. Both pieces present a building set upon a sort of base, almost a raised socle, that includes the area of the portico. The model from the Heraion of Argos, with its rectilinear walls and square-section pilasters, perhaps reflects the use of mud-brick walls and wooden vertical supports in contemporary full-scale buildings. In contrast, the apsidal shape of the model from Perachora suggests walls made of irregularly cut stone, at least for the socle, and trimmed trunks for columns. Along the outer walls of both models, the open windows evidently reproduce the means for illumination and ventilation of the interiors of actual buildings; their triangular shape, heir to a long tradition (including the openings in the great Mycenaean *tholoi*), was meant to serve static needs and avoid the use of wooden architraves, often subject to warping, in bearing walls. The form of the roof was given particular attention in both models. The two different shapes have suggested that there were two different methods of roof carpentry, but both were probably associated with a roof made of plant material, either branches or rushes. The building reproduced in the Corinthian example may have had a ridge pole that rested on a pillar located at the center of the apsidal space, accompanied by a series of oblique rafters rising in a fan shape; in the source for the Argive example, the roof may have been supported by a framework of perpendicular beams arranged to correspond to the rectilinear contours of the walls.

Due to the religious context in which these models were found, and the attention their makers obviously gave to rendering detail and decoration, it was long believed that the objects were in fact miniature depictions of the temples built for the sanctuaries. Recently, however, several doubts have arisen about this supposition. Rather than reproductions of cult buildings, the models are now believed by some scholars to be symbolic representations of a home; this conclusion is based upon reference to another architectural model whose function has been established by archaeological research. Thus these small masterpieces were likely produced as offerings to Hera in her role as the divine protector of the domestic setting, although the possibility cannot be excluded that the models symbolized instead the "home of the divinity."

MAKING WOODEN ARCHITECTURE INTO STONE, STANDARDIZATION OF BUILDING MATERIALS, AND THE "REBIRTH" OF TILE IN GREEK ARCHITECTURE: THE TEMPLE OF APOLLO AT CORINTH AND THE TEMPLE OF POSEIDON AT ISTHMIA

The passage from the eighth to the seventh centuries B.C. was marked by two outstanding innovations in building technology: the use of stone as the principal building material and the appearance of terracotta tiles for the roofs of buildings.

In reality, the use of clay elements for roof coverings is part of a very ancient tradition, dating to the fifth millennium B.C. Excavations at the site of Lerna in the eastern Peloponnesus have revealed the remains of a large rectangular structure (12 by 25 meters), dated to the third millennium B.C., which archaeologists named the House of the Tiles precisely because of the discovery of a large number of stone and terracotta elements (each measuring about 22 by 26 centimeters) associated with its roof. However, with the destruction of the Mycenaean palaces and the loss of related technical knowledge, the system of making roofs with clay elements—using flat tiles with a trapezoidal shape and semiconical covers—was abandoned in favor of roofs made with both plant and stone materials directly available in nature.

The reappearance of terracotta tiles in architecture during the first half of the seventh century B.C. was one of the many cultural transformations that marked the transition from the Geometric period to the early Archaic. Written in the first century A.D., Pliny

Top. Drawings of two architectural models:
A. From Perachora and
B. From Argos.

Bottom. Hypothetical reconstructions of the building systems used in structures depicted in votive models.

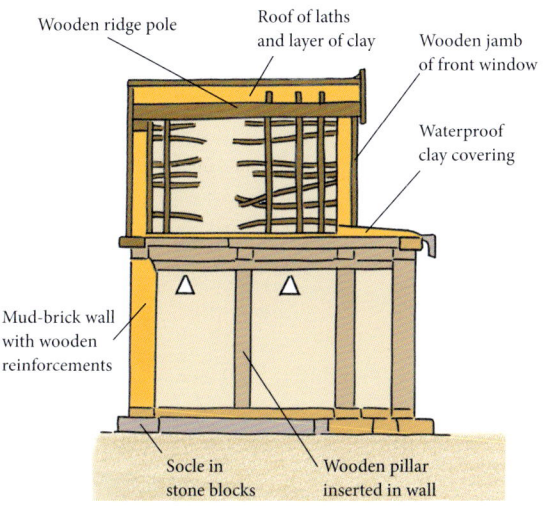 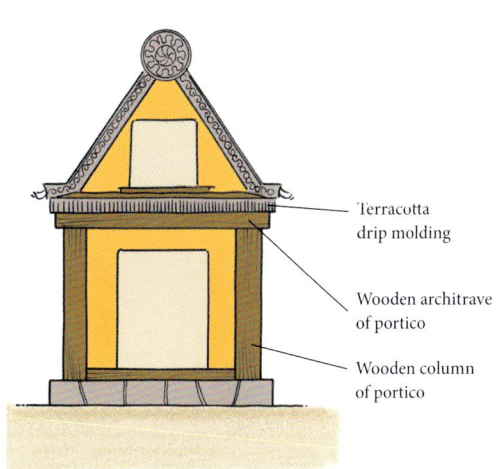

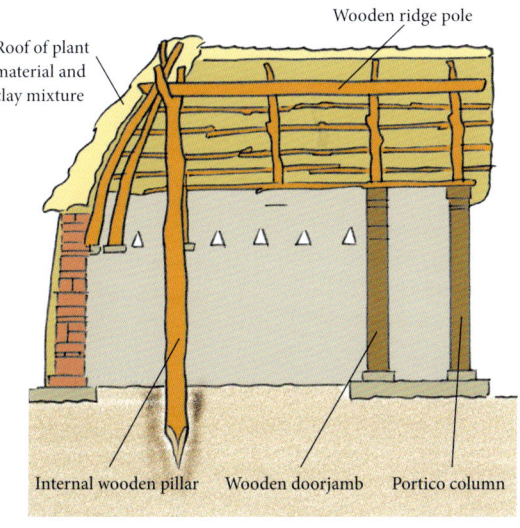 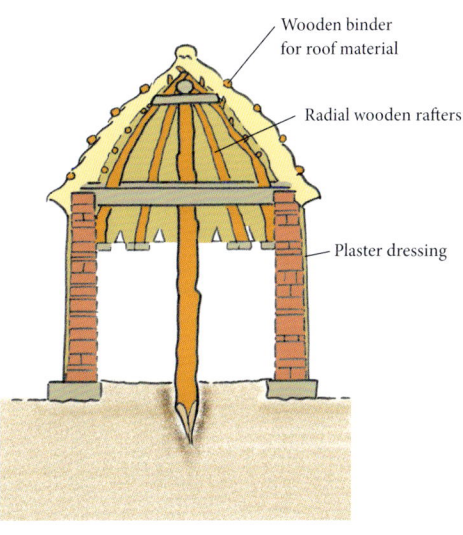

CONSTRUCTION TECHNIQUES IN THE GREEK WORLD 83

Archaeological excavations have uncovered large portions of ancient Corinth. The map below shows the topography of this Greek and Roman city, overlaid with indications of the outline of the modern city. At the center, in the northwestern sector of the agora, stands the hill on which the Temple of Apollo was built.

the Elder's encyclopedic work *Natural History* attributes the beginning of the production of tiles to the Cypriot context (7.195): "Tiles were invented by Cinyra, son of Agriopa, as well as mining for copper, both in the island of Cyprus." This passage acquired particular significance to archaeologists following the discovery at the site of Amathous, on the southern coast of the island, of several tiles that may have been made in the seventh century B.C. Although this date has not been solidly confirmed, scholars nevertheless agree in assigning to Corinth, and to Corinthia in general, a central role both in the transformation of monumental architecture into stone forms, and in the rebirth, after an interruption of about five centuries, of the production of clay elements for making roofs, an artisan activity favored in the area by the presence of abundant quarries of excellent clay.

Between 675 and 650 B.C. the first large temple to Apollo was built at Corinth, the homeland of many Greek colonies in the West. Unfortunately, little is known of the building's shape because of its total reconstruction following a fire around 550 B.C. It was probably a temple of the distyle *in antis* plan, with a pronaos but not a peristyle, and dimensions of roughly 11 by 33 meters. The numerous architectural elements attributed to this construction reveal the adoption of a building technique already dependent on the use of squared stone. The builders employed oolitic limestone, soft and fine grained, that was quarried from the same bedrock on which the city was then being built, ensuring that material was readily available to the work site without the need for extended transportation. Archaeological evidence has revealed that this limestone was used in the construction of the temple for the foundations as well as for the socle of the walls. The socle was constructed using well-squared blocks, and its framework was skillfully reinforced by the insertion of connecting clamps between the cornerstones. Above the socle stood a wall, perhaps built primarily of mud bricks but at the top of which was a row of stone blocks—representing their first known usage in this context—decorated with plant motifs in red and black. The primary purpose of these blocks was to secure, by means of grooves on their upper surface, a series of wooden elements that supported the rafters of the roof.

This experiment in the use of stone as a principal building material is of particular interest because Corinth was a city known for its skills in the construction of wooden ships. Perhaps it was master ship carpenters who discovered (or rather rediscovered) that limestone could be worked with the same tools as those used for wood: tools of direct percussion, such as the axe, mallet, and hatchet, as well as tools of indirect percussion, including chisels.

The Corinthian builders of the Temple of Apollo used wood for more than just the framework of the roof. Grooves in both faces of the socle blocks of the walls represent traces of a wooden frame. The vertical boards of this frame were anchored to the walls with wooden nails placed at regular distances, and connected to one another by means of additional boards that ran horizontally along the bottom and top of the wall. This system appears to continue the ancient tradition of associating wooden supports with the masonry structure of walls, in order to increase the latter's capacity to support the roof; although mud construction cannot be confirmed, this support would have been a particular requirement in those cases where temple walls were still being made in part from mud bricks. We do know for certain, however, that the walls were covered on both sides from the socle up with a thick layer of plaster, which may have been painted.

Archaeologists have proposed that the temple had a pitched roof with three or four slightly inclined slopes each approximately 9.5 degrees. The covering was composed of terracotta tiles of the heavy-combination type, in which a slightly concave tile was joined to a convex junction cover, creating a unit that measured roughly 67 centimeters on each side. This system, conventionally referred to as Proto-Corinthian, required the creation of individualized pieces, each of which was designed and produced on the basis of its location on the roof; these elements included tiles situated along the ridge line and in the projecting areas between the slopes, as well as cover tiles that were characterized by a trapezoidal termination and placed along the eaves. It is possible that the cream-yellow color of these tiles, when placed in conjunction with the architectural elements that were painted black, served to create a polychrome design with a decorative purpose, the pale vertical bands alternating with dark bands accented in red.

During the same decades as construction of the Temple to Apollo at Corinth, the first peripteral temple dedicated to the cult of Poseidon Isthmios ("of the Isthmus") was erected at the nearby sanctuary of Isthmia on the isthmus of Corinth, site of the spring celebrations that became Pan-Hellenic in the sixth century B.C. This large building, 14.40 by 39.25 meters, had as its central body a *hekatompedon* ("hundred feet"), with an almost square pronaos and an elongated cella. The full length of both spaces was divided by an axial row of wooden columns that supported the roof rafters. As with the Temple of Apollo, at least part of the building was made in blocks of oolithic limestone, the local availability of which greatly accelerated preparation of architectural pieces. The recourse to material relatively easy to acquire was just one carefully considered element at a well-organized

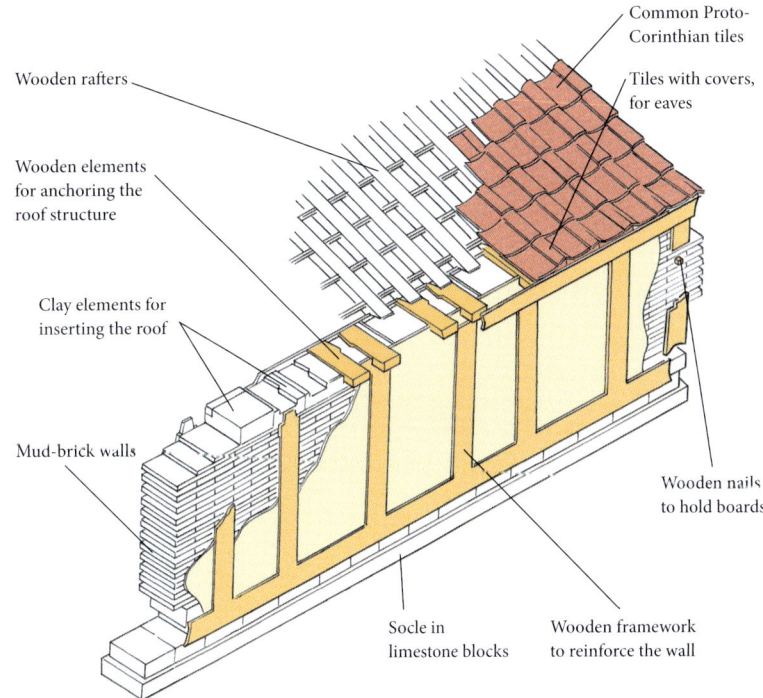

Top. The Temple of Apollo at Corinth was made using a sophisticated method of construction. On a socle built with limestone blocks stood walls that were probably made of mud bricks, reinforced by the insertion of horizontal and vertical wooden elements, and covered with plaster.

Following page. The face of an ancient limestone quarry near Isthmia and Corinth.

Opposite, top. Schematic re-creations of the Temple of Poseidon at Isthmia.

Opposite, bottom. System of roofing used in the Temple of Poseidon.

site whose workers possessed advanced technical skills and where the sizes of building materials had been standardized. Limestone, for example, was removed from the quarry in blocks that had been cut to nearly their final size, thus avoiding transportation of stone destined to be eliminated later. When the limestone blocks arrived on site, each piece was finished by stonecutters to obtain square ashlars (known as bonding blocks, or diatons), measuring roughly 27 centimeters high and as wide as the thickness of the walls, and readily intalled by a stonemason. The difference in size between the temple's two primary areas required the creation of two types of blocks: those destined for the walls of the cella, about 55 centimeters wide, and those assigned to the walls of the pronaos, about 65 centimeters wide. The average length of the ashlars was 82.5 centimeters. These measurements point to another factor that lay behind the building process for the temple at Isthmia—the desire for a precise relationship among the length, width, and height of the blocks of the cella walls, corresponding to a ratio of 3:2:1.

As with the Temple of Apollo at Corinth, the walls of the structure at Isthmia were associated with a series of wooden elements designed to assist in support of the roof. At Isthmia there were pillars, 32 to 37 centimeters wide, set against the external face of the walls of the central body, probably on the axis of the columns of the peristyle. In this building the roof was again of the type with three or four slopes, but it rested not just on the walls of the cella and the pronaos but also on the wooden architraves of the peristyle and on the axial colonnade. This arrangement resulted in a different structure for the interior carpentry, with the insertion of a horizontal frame of beams upon which rested a longitudinal system of additional beams and transverse rafters.

Undoubtedly, the principal element that the building at Isthmia had in common with the Temple of Apollo at Corinth was the Proto-Corinthian typology of the roof. The tiles used measured about 65 by 69 centimeters, with a thickness of between 4 and 5 centimeters; each element weighed approximately 30 kilograms, for a total weight of 53,000 kilograms. The shape of the tiles was similar to that seen at Corinth, with the addition of a characteristic triangular piece at the center of the tiles located on the eaves.

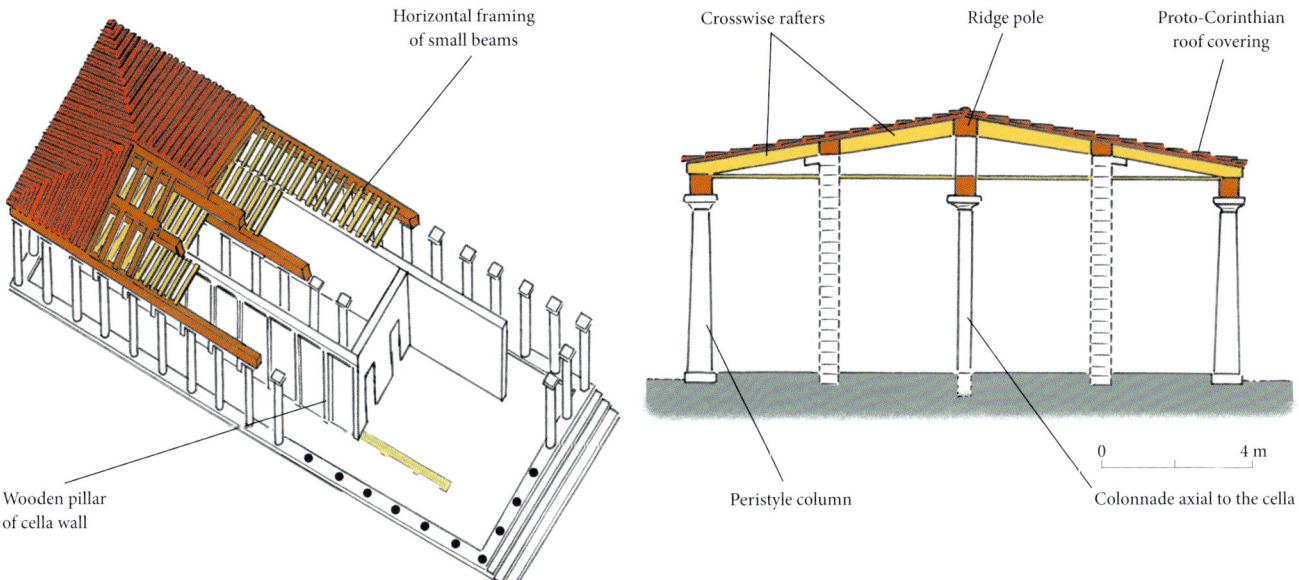

Tiles of the Proto-Corinthian type were also found at another site within the territory of Corinth, in the sacred area of Perachora discussed above. These tiles were from the roof of a building erected during the seventh century B.C. Archaeologists have not ruled out the possibility that this structure was the Temple of Hera Akraia, rebuilt after the collapse of an ancient apsidal structure dating to the Geometric period. At Perachora, unfortunately, the archaeological evidence is extremely spotty due to subsequent renovation of the temple in the sixth century B.C. Tiles of the same type have also been identified in the sanctuary of Apollo at Delphi, where Corinthians of the second half of the seventh century B.C. had offered a *thesaurus* ("treasury"). The Roman historian Plutarch (*Oracles at Delphi* 13), attributed construction to the tyrant Cypselus: "Sarapion asked the guides why it is that they call the treasure-house, not the house of Cypselus the donor, but the house of the Corinthians.... We heard them say earlier that when the despotism was overthrown, the Corinthians wished to inscribe ... the treasure-house here with the name of their city." The building, measuring

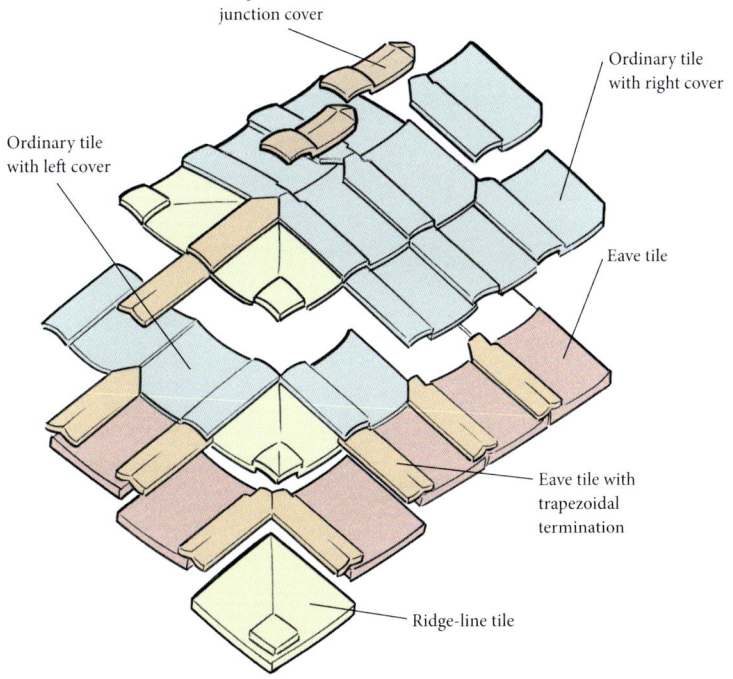

5.8 by 13.2 meters and perhaps Doric distyle *in antis*, was probably made by importing from Corinth not only the roof tiles but all the elements required for the work site, from stone materials to artisans.

CONSTRUCTION TECHNIQUES IN THE GREEK WORLD 87

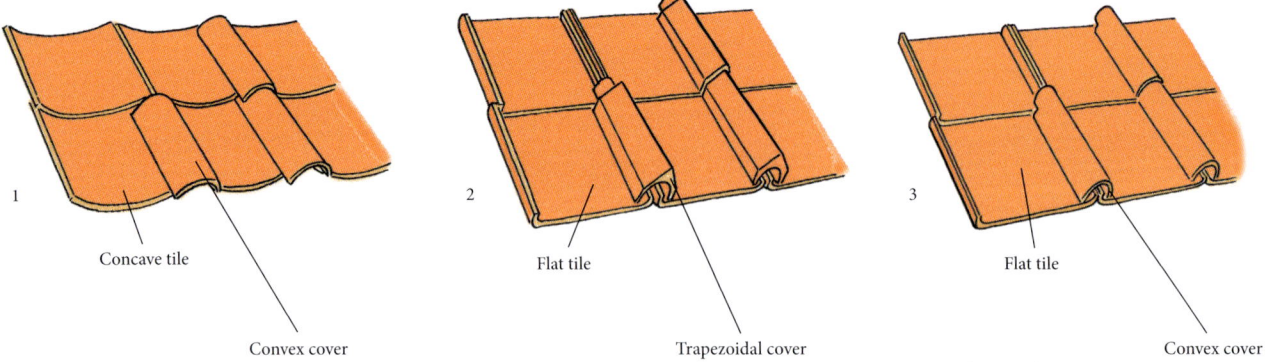

Principal types of roofing tiles and covers: 1. Laconian roof; 2. Corinthian roof; 3. Hybrid roof.

Opposite, top. The Temple of Apollo at Thermon in Aetolia represents one of the most ancient uses of a Doric frieze with metopes and triglyphs.

Opposite, bottom. Stratification of the area of the Temple of Apollo at Thermon. Violet: Megaron B (Protogeometric period); blue: arrangement of stone slabs (late eighth–early seventh century B.C.); green: the peripteral temple (late seventh century B.C.).

THE TEMPLE OF ARTEMIS AT SPARTA AND THE "LACONIAN" ROOF

Perhaps as early as the ninth century B.C., in the southeastern area of inhabited Sparta near the Eurotas River, a sanctuary of significant size arose dedicated to the cult of Artemis Orthia ("of the Dawn"). Located in one of the most sacred areas of the city, it was the site not only of musical contests and recitations but also of the ritual flogging of young Spartan men, a ceremony that was carried out to establish their manhood, as described by Plutarch (*Ancient Customs of the Spartans* 239): "The boys in Sparta were lashed with whips during the entire day at the altar of Artemis Orthia, frequently to the point of death, and they bravely endured this, cheerful and proud, vying with one another for the supremacy as to which one of them could endure being beaten for the longer time and the greater number of blows. And the one who was victorious was held in special repute. This competition is called 'The Flagellation' and it takes place each year."

The association of a variety of rituals at this site led to the creation of a composite sanctuary, where during the Hellenistic period a cult building and an altar were located together with a great theatrical cavea, or auditorium with seating for spectators. The earliest architectural structure in the sacred area appears to date to the ninth century B.C., when a rectangular stone altar and the oldest known peribolos wall were erected on terrain that had been carefully paved in stone. The temple was not built until near the end of the following century, as part of a process of general renewal that included creation of another altar and perhaps enlargement of the temenos. Little of the building's structure has survived, as it was replaced by a new temple early in the sixth century B.C. The earlier building was probably rectangular (4.5 by 12 meters), lacking an exterior peristyle, with a cella divided in half by a row of axial columns. A socle in blocks was set on top of a foundation of river rocks, while the walls were likely made of mud bricks, reinforced by a series of small wooden pillars set against the internal faces of the walls and probably aligned with the internal colonnade.

Replacement of this temple's original roof was among the changes made to the building during the seventh century B.C. The new roof was built using terracotta elements that were composed of large, slightly concave tiles, together with convex cover tiles that were produced separately. This type of roof, referred to as Laconian, also required the use of specially crafted pieces of tile for placement along the ridge line, where a series of larger covers extended to the facade (which featured a large disk-shaped acroterion), as well as along the eaves, where the covers ended with semicircular polychrome antefixae.

THE SANCTUARY OF THERMON AND THE FIRST USES OF TERRACOTTA FOR TEMPLE DECORATION

At Thermon, on a small plateau that is located along the slope of Mount Mega Lakkos in Aetolia, excavations of the Sanctuary of Apollo Thermios (a divinity of heat and protector of flocks) have made possible reconstruction of the area's history extending from the Bronze Age to the Hellenistic period. The oldest phase at the site has retained substantial traces of the construction of various forms that were grouped around a large structure with an apsidal form, known as Building A, which measured 22 meters in length and has been dated to sometime in the Mycenaean period. To the south of Building A is the so-called Megaron B (21.4 by 7.3 meters), which was erected over the course of the eleventh century B.C., and is a tripartite structure with slightly curvilinear longitudinal walls. Roughly a century after the destruction of Megaron B at the end of the ninth century B.C., eighteen stone slabs were arranged around its remains, but at a higher level, in a fork-shaped configuration. These slabs were apparently bases for wooden poles, and provide evidence of an installation that archaeologists have interpreted either as an open-air enclosure, or as reinforcement for a mud-brick wall that formed part of a building made without stone foundations and that consequently has left no further traces.

During the last decades of the seventh century B.C. at Thermon, building practices that were still distinguished by the use of perishable materials evolved to those methods necessary for construction of a true peripteral temple—as represented by the structure known as Temple C—with five columns on each end and fifteen along each side wall. The building measured 12.13 by 38.32 meters and was reconstructed at the end of the third century B.C. according to the same dimensions—in part by reusing architectural elements—with a long cella divided in two aisles by means of an axial colonnade and an almost

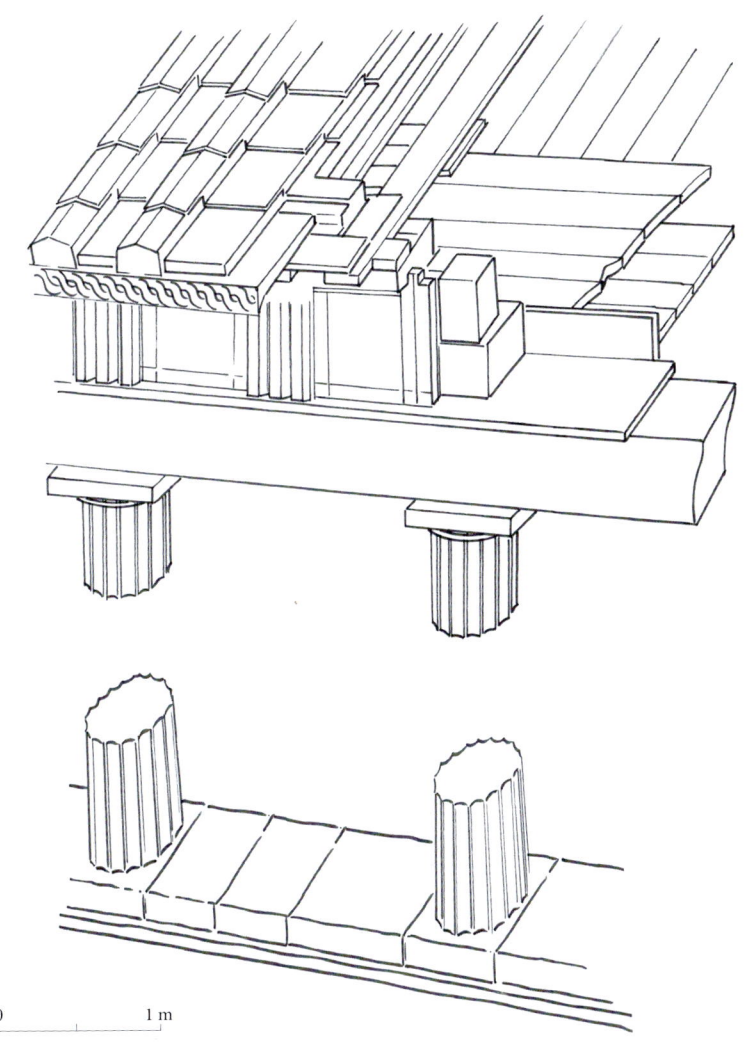

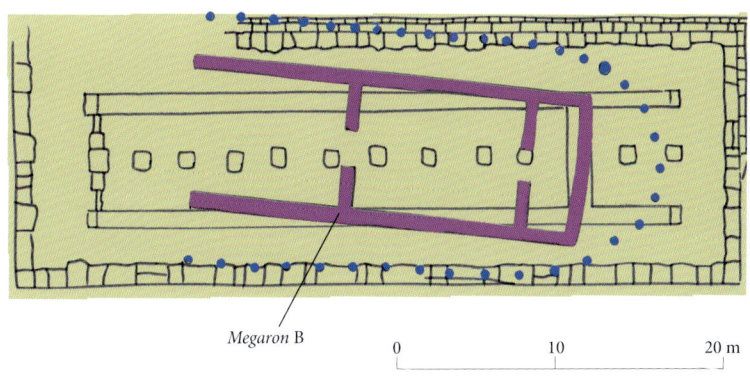

Megaron B

At Olympia the Altis (as the Greeks called it) was the most important sanctuary in Hellenism, especially following the institution of the Olympic Games in 776 B.C. This image shows the remains of the great temple attributed to the cult of Hera, erected around 600 B.C., in the northern sector of the sacred area.

square opisthodomos. The building technique combined the stone elements of a solid base with walls that were still primarily made with perishable materials. The walls, in fact, were made of mud bricks, while the column shafts, along with the trabeation and structural work of the roof, were made of wood.

Archaeological excavations at the end of the nineteenth century uncovered a large collection of terracotta elements related to the temple's tile roof and decoration. Analysis of the shape of the tiles allowed reconstruction of the roof, which was arranged with three slopes and a pediment on the front face. The original mantle—later replaced along with the wooden columns of the peristyle during the second half of the sixth century B.C.—was composed of flat tiles nailed to a wooden frame, and pentagonal covers that ended, along the line of the eaves, in antefixae decorated with polychrome female faces. Rainwater was drained by way of downspouts, decorated with the heads of lions, that were inserted at the end of the roof in the angles formed by the ridge lines between the eaves.

Certainly the most interesting archaeological evidence at Temple C relates to its trabeation, which included several large terracotta slabs (measuring roughly 99 by 87 centimeters) with polychrome decoration in black, white, and red. Depicting mythological scenes (including Gorgons pursuing Perseus, and Aedon and Chelidonis killing Itys), the paintings were created before the terracotta pieces were fired but after the design had been incised on the still damp slab with a punch. Given that these pieces were produced to cover a traditional trabeation made of wood, the method of installation remains the subject of debate.

The clay decoration of the frieze of Temple C at the Sanctuary of Apollo Thermios does not represent a unique case in central Aetolia. Indeed, recent studies have called attention to another temple, erected in the mid-seventh century B.C. just northwest of Temple C and perhaps dedicated to the cult of Artemis, that was given not only rich polychrome terracotta decoration on its roof but also a Doric frieze in clay elements. A fragment of a triglyph

from this frieze, decorated in brown with white glyphs, preserves part of the slot for the side metope. Toward the end of the century the construction of the Temple of Apollo Lyseios at Thermos also included the creation of a Doric frieze in clay elements, for which the painted metopes (each measuring about 60 by 90 centimeters) have been found.

THE CONSTRUCTION SITE FOR THE HERAION AT OLYMPIA

The most important sanctuary of Greek culture arose at the foot of the Hill of Kronos, between the Alpheios and Kladeos rivers in southern Elis, in an area inhabited since the beginning of the second millennium B.C. This location, which the Greeks called Altis (a variation on *alsos*, meaning "sacred wood"), became in 776 B.C. the site of the famous Olympic games, an event that drew the entire Greek world, including the colonies, every four years.

An important process of monumentalizing began at the sanctuary around 600 B.C. with the construction of a large temple dedicated to the cult of Hera. Pausanias (*Descr. of Greece* 5.16.1) visited the building in the second century A.D. and offered these details of the temple's history and form: "The Elean account says that it was the people of Scillus, one of the cities in Triphylia, who built the temple about eight years after Oxylus came to the throne of Elis. The style of the temple is Doric, and pillars stand all around it. In the rear chamber one of the two pillars is of oak. The length of the temple is one hundred and sixty-nine feet, the breadth sixty-three feet, the height not short of fifty feet. Who the architect was they do not relate." In reality, the temple that Pausanias saw did not result from a concentrated building program, but rather from a long series of undertakings that took place over a period of more than seven centuries. At the time Pausanias visited the temple, it differed greatly from the original building constructed at the beginning of the sixth century B.C.

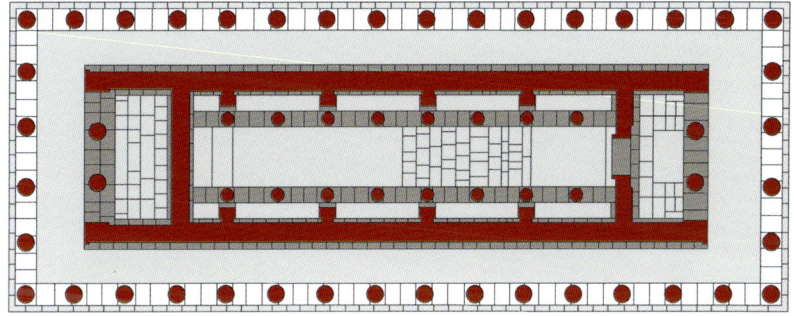

Facing east, the temple (18.76 by 50.01 meters) included a peristyle of six by sixteen columns, set upon a base with a single step. The internal arrangement had a distyle *in antis* pronaos, and the cella was divided into three aisles by two rows of eight columns, perhaps arranged in two superimposed orders; behind the cella was the opisthodomos, also distyle *in antis*. The walls stood on a socle made of carefully hewn stone, while the upper part of each wall was mud brick. A desire to free the internal space of the cella from an axial colonnade, which would have had a negative effect on the central location and visibility of the cult statue, led the architect to devise an ingenious composite solution. The two side aisles, narrower than the aisle in the middle, were articulated along the walls by a succession of large pilasters that alternated with the internal columns. This arrangement produced two opposing series of large niches for the display of votive offerings; it also made possible a reduction in the distance between the two longitudinal walls, and thus a decrease in the length of the roof rafters. Elimination of the axial colonnade therefore made possible the creation of a wide central aisle, at the far end of which was placed the cult statue of the goddess.

Excavation of the monument at Olympia and analysis of its architectural remains have allowed the reconstruction of many aspects related to the temple's construction, beginning

The building attributed to the cult of Hera in the sanctuary at Olympia was a peripteral temple, with six columns on the short sides and sixteen on the long ones. Within, it had a pronaos and a distyle in antis *opisthodomos, while the cella had long walls with pilasters that alternated with the internal colonnades.*

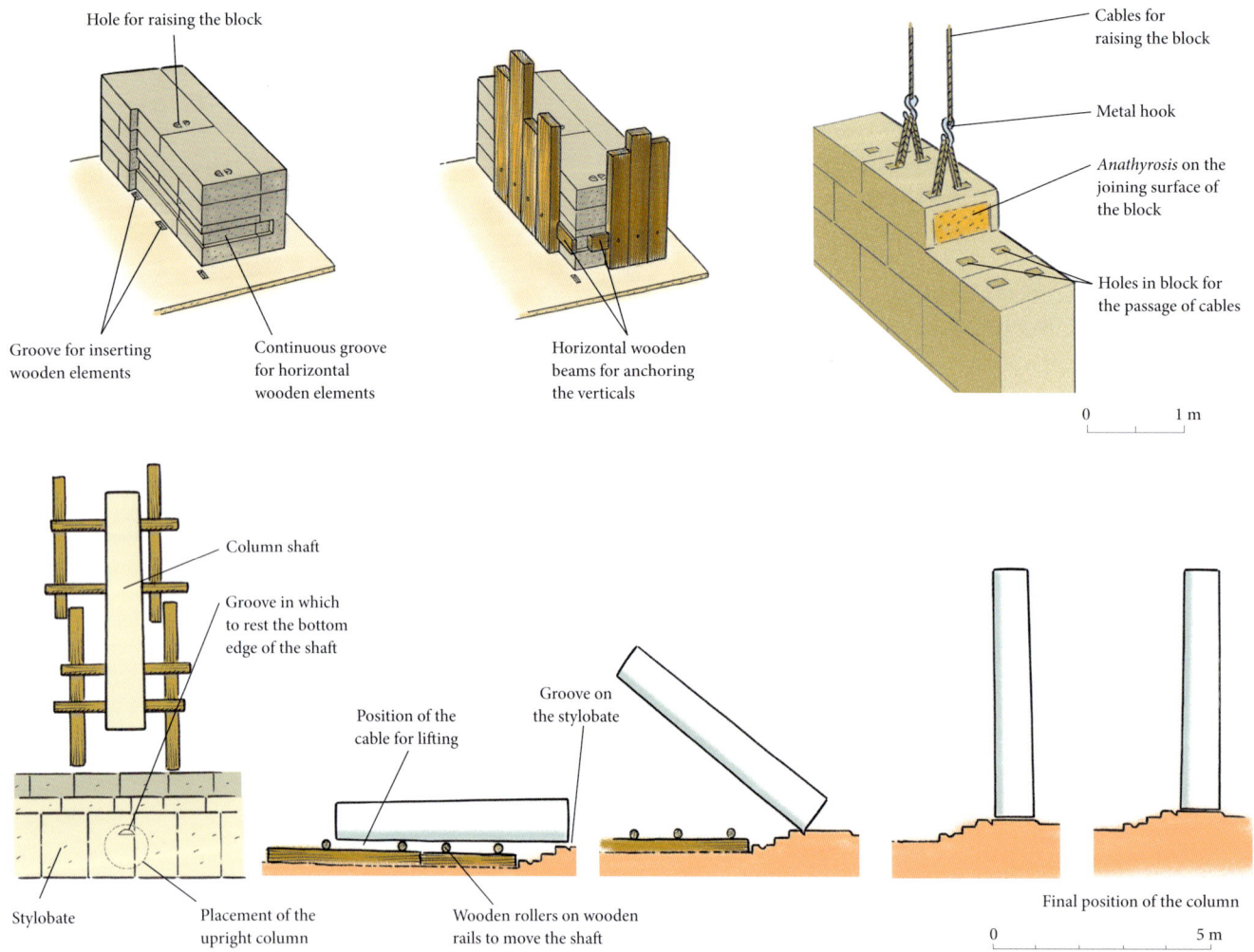

Top, left and center. Method of placing a wooden cover on the antae of the pronaos and the opisthodomos of the Temple of Hera at Olympia.

Top, right. System of raising blocks followed in construction of the Temple of Hera.

Above. System of raising the monolithic column shafts used in the temple.

with preparation of the foundations. It was first necessary for the workers to make a deep cut in the rock in order to position the temple at the northwest corner of the sanctuary, at the foot of the Gaion (which represents the extreme extension of the Hill of Kronus). The next step involved the creation of foundations that were designed to reinforce the building's position on the terrain. Deep trenches were dug, at the bottom of which the architect ordered a bed of limestone and sandstone rubble that served not only as a drainage layer but also as the impost for several courses of blocks of natural conglomerate. All other stone parts of the construction, beginning with the uppermost course (Gr. *euthynteria*), which marked the end of the foundations above ground, were made using conchiferous limestone, a material that is strong yet easy to work into square elements. Its level of workability made possible the adoption of certain technical solutions that facilitated the installation of the various architectural elements. The blocks of the socle were chiseled, most notably along the upper faces, to form a pair of grooves that served as continuous canals for the passage of ropes used to raise each block and position it inside the structure.

The socle of the longitudinal walls reveals a high level of skill at construction. Within this section of the walls, blocks of diverse sizes were installed in various ways. Perhaps to provide greater stability to the large loads of the roof that would be placed over the rooms, the internal faces of the walls were made with four courses of limestone blocks, each of which was placed with one of its larger faces positioned toward the interior of the room. In contrast, the external faces of the walls were made with blocks that were arranged in a single vertical row, rising to a height equal to that of the four internal courses; this strategy was probably designed to emphasize the structure's monumental character. The technique clearly had significance as it reappears in the rear wall of the opisthodomos and in the two masonry wings that flank the entrance to the cella. Contacts between the blocks were executed by means of sharp junctures, without recourse to any system of metal clamps.

As observed earlier, a wall in mud bricks rested above the socle. Here, as well, the builders adopted several skillful technical methods, protecting and reinforcing the weaker areas of the walls (which corresponded to the antae of the pronaos of the opisthodomos) with a covering of wooden elements that acted with the walls to support the roof.

Since no remains of trabeation have been found, it can be conjectured that the structure was made in large part of wood, to which were nailed bronze metopes with embossed decorations. The roof of the building was constructed with clay elements of the Laconic type, using slightly concave tiles and convex covers; the latter, corresponding to the line of the eaves, ended in antefixae decorated with a twelve-petaled flower. A large circular acroterion stood at the center of the roof; it also was made of terracotta, but with the addition of polychrome decoration, and extended to a diameter of about 2 meters. The difficulties in modeling, decorating, and firing such a large piece again testify to the high level of technical expertise that had by then been attained in the field of terracotta. One of the specialized strategies adopted was that of leaving an open space at the center of the acroterion, to absorb any variations in the volume of the clay during firing and thus avoid the formation of internal cracks.

Pausanias remarked, as we have noted, that in his time one of the two columns of the opisthodomos of the Heraion at Olympia was made of oak. The comment is of particular interest, first of all for its identification of the type of wood used. Oak (Gr. *drys*) was widely employed in ancient architecture, although Theophrastus, in his botanical history (*Enquiry into Plants* 5.5.1), considered the material difficult to work. Pausanias's remark is also significant because it helps to explain certain anomalies in the construction of the temple that came to light during excavations: the design of the shafts of the temple's Doric columns differed from that of the capitals, and not all the limestone originated at a single quarry. In fact, monolithic shafts flanked others that were made by stacking a series of drums, while columns with capitals that featured a very flattened profile, typical of the Archaic period, flanked others with a more expanded echinus, dating to the Classical or Hellenistic

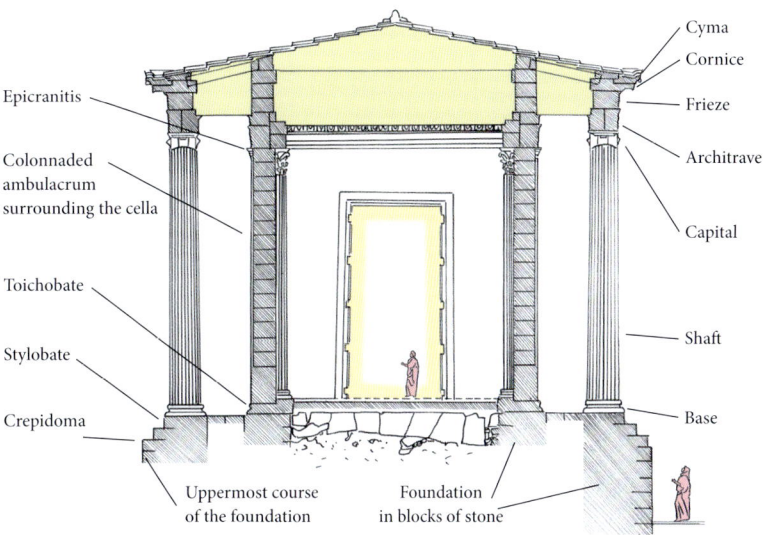

Cross section of the Temple of Latona (Leto) at Xanthos in Lycia, illustrating some of the principal elements of Greek temple architecture. The building was constructed at the end of the first half of the second century B.C. in the sanctuary dedicated to the mother of the twins Apollo and Artemis.

CONSTRUCTION TECHNIQUES IN THE GREEK WORLD 93

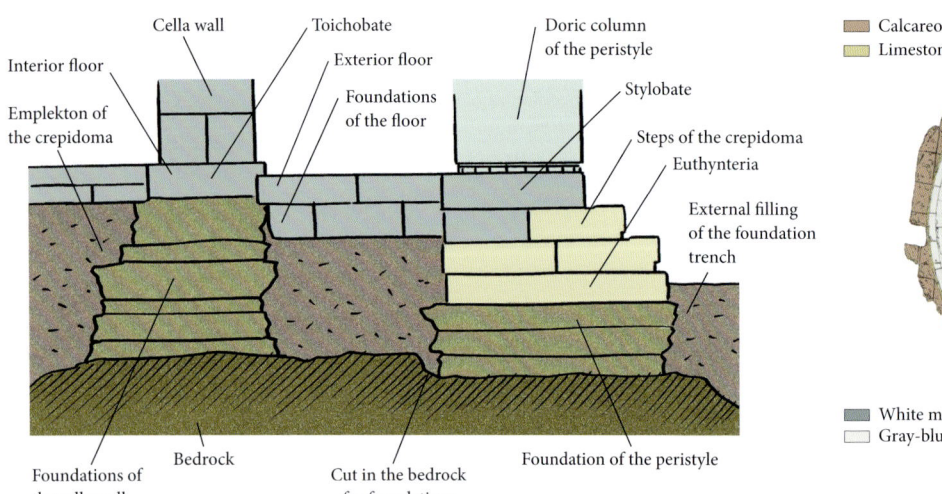
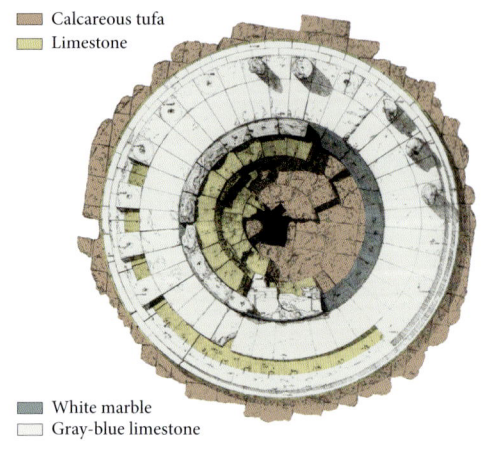

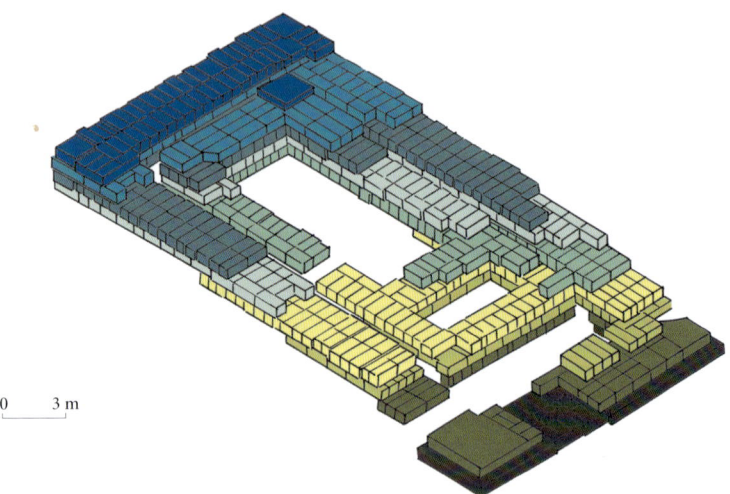

Top, left. Arrangement of the foundations of the Great Temple at the Sanctuary of Apollo on Delos, begun around 475 B.C.

Above. Organization of the foundations in the Temple of Asclepius (early fourth century B.C.) at Gortys in Arcadia, which called for overlaying numerous rows of squared blocks in an alternating arrangement.

period. The grooves, or fluting, in the shafts also differed, varying in number from sixteen to twenty. The explanation for this array is that the columns were progressively replaced. All the originals (5.21 meters high, about 1 meter in diameter) were made of wood, but over the course of centuries they were replaced by columns whose proportions and capitals reflected those in fashion at the time of the change. The stylobate (Gr. *stylobates*) still bears traces of the original wooden columns, along with several shallow grooves used for positioning the monolithic stone columns, including triangular sections indicating where the bottom edge of each shaft rested as it was raised.

FOUNDATIONS AND BASES OF BUILDINGS

The construction of every building began with preparation of the ground for the creation of foundations (Gr. *themelion*), which served to take the entire weight of the structure, divide its loads, and transmit them to the ground.

In general, this process began with digging a trench or ditch (Gr. *orygma* or *orychē*) deep enough to reach a solid layer that would bear the weight of the loads. It sometimes happened that removal of the surface layer (Lat. *humus*) revealed bedrock; in such cases the next step was to make a series of cuts (Gr. *entomē*) in the stone, in order to level its surface and create a uniform plane on which to position the building. This technique was used, for example, at the work site of the Temple of Hera Lacinia near Croton, which was erected in the second quarter of the fifth century after the destruction of the Archaic building, and also at the site of the Temple of the Dioscuri at Agrigento, which was built during the same period.

Foundation systems were of three types. Continuous foundations featured structures that joined all the bearing parts of a building from below, without interruption. Discontinuous foundations supported isolated pillars that corresponded to a building's bearing elements. A bed foundation, used where the terrain

presented features of instability, involved the entire surface of a monument, as at the Tholos of the Sanctuary of Athena at Delphi.

For more-modest constructions, foundations were usually made of irregularly shaped rubble arranged in roughly continuous and horizontal rows; the work sites of foundations for large structures were built of several courses of squared blocks installed without mortar. The latter operation involved digging a trench that usually projected beyond the width of the structure to be erected, thus allowing the space necessary for the movement of workers. Beginning with the bedrock or lowest level and extending to the grade plane of the structure, individual courses were laid out beyond the perimeter of the structure to be built above them, thus increasing the area of the lower surface that contacted the ground and resulting in improved distribution of the loads.

Because foundations were destined to remain underground, the stone used was generally inferior in quality to that for above-ground structures, and individual pieces were worked in a less careful fashion. Moreover, architectural elements from previous constructions were often reused in the foundations. For example, the basement of the treasury of the Sikyonians at Delphi, which was built at the end of the sixth century B.C., reused much of the framework of a preceding monopteros and a tholos.

Unlike the other foundation courses the euthynteria—the layer that formed the regular plane on which the walls would stand—could

Top. Close-up view of the crepidoma of the Propylaea of the Acropolis in Athens (437–432 B.C.). Note the choice of Pentelic marble for the euthynteria of the central body, and the gray-blue stone of Eleusis for that of the Pinacoteca.

Opposite, top right. Plan of the tholos in the Sanctuary of Athena at Delphi, with indications of the different building materials used.

CONSTRUCTION TECHNIQUES IN THE GREEK WORLD 95

The state of preservation of the Temple of the Dioscuri (460–450 B.C.) at Agrigento makes it possible to visualize the system followed by workers to cut the building's supporting stereobate out of bedrock.

itself extend above ground. For this reason it assumed a certain significance and was at times made of a more durable and more workable type of stone. The Propylaea designed by Mnesicles to monumentalize the principal gateway to the Athenian Acropolis clearly illustrates the importance that Greek architects attributed to the euthynteria. For the central body of the complex, designed with a large prostyle hexastyle front, Mnesicles chose to make the euthynteria with the same Pentelic marble used for the crepidoma (Gr. *krepis*) and walls; for the two wings, however, which he wanted to function as a frame, he used a contrasting color, creating a course of blocks in gray-blue limestone from Eleusis.

Immediately above the euthynteria was the true socle of the building, the crepidoma. This element was comprised of a series of steps, at the top of which could stand either the colonnade of a hypostyle facade or one of the building's walls. The crepidoma usually had three steps, but the number varied from just a single step (as at the Temple of Hera at Olympia) to seven (at the Temple of Apollo at Didyma). The Temple of Hera at Agrigento, erected in the second half of the fifth century B.C. at the eastern end of the well-known Valley of Temples, had a crepidoma with four steps, the lowest of which was decorated with horizontal grooves marked off at right angles to correspond to the joints. At other buildings, such as the Temple of Apollo at Bassae (dated to the fifth century B.C.) or in the peripteral Temple of Asclepius at Kos (second century B.C.), the same type of decoration of the crepidoma steps was further enriched by the insertion of a molding of a reversed cyma.

The final step of a crepidoma assumed special meaning, and thus it was usually built to be higher than the other steps. On its upper surface rested the columns and walls of the building, which assumed, respectively, the names stylobate and toichobate (Gr. *toichobates*).

GREEK WALLS

Vitruvius (2.8.5) praises the Greek world's tradition of building in stone: "The walling of the Greeks is not to be made light of.... When they depart from ashlar, they lay courses of lava or hard stone, and, as with brick buildings, they bind their joints in alternate courses, and so they produce strength firm enough to last." He thus introduces a distinction between two types of Greek construction techniques, based specifically on the use or absence of squared stones. In fact, the Greeks employed numerous methods for the construction of walls, with the choice depending on the static behavior of the framework inside the building as well as on the availability and workability of local material.

In terms of sheer monumentality, the most impressive type of wall was without doubt the Cyclopean, used in the powerful terracing walls of the Archaic-period Temple of Hera in the sanctuary at Argos. Irregularly shaped blocks were installed without mortar, creating numerous discontinuities among the elements of the wall. The empty spaces resulted in structural instability that was addressed by trimming the faces of the blocks to create more precise contact among the elements, and by filling the interstices with smaller stones.

The development of Cyclopean walls represents the birth of polygonal masonry, composed of stone blocks dressed with relatively straight sides, so that on the face of the wall each block assumed the shape of a polygon. In general, each block had more than four faces of juncture; according to the level of dressing, the technique produced either irregular or regular polygonal masonry.

Constituting an intermediate stage between Cyclopean and regular polygonal masonry, irregular polygonal masonry was composed of stones only roughly cut, so that when blocks were set in place it remained necessary to use small stones as fill for the interstices in the

Top, left. The terracing wall in Lesbian-style polygonal masonry located to the east of the temple in tufa in the Sanctuary of Athena at Delphi.

Top, right. Tower in regular opus quadratum *of the fortress of Aigosthena, erected in the fourth century* B.C. *in northwestern Attica.*

Top, left. Wall in gneiss blocks of a house from the Hellenistic period located along the Theater Road on Delos. Note the presence of a window framed in white marble.

Top, right. One of the trapezoidal-masonry towers of the fortress of Eleutherai, built in the fourth century B.C.

wall. There are numerous examples of this technique in Greek structures, primarily for terracing walls or city walls, such as the Hellenistic fortifications at Oiniades in Acarnania.

Regular polygonal masonry called for greater effort in the dressing of the individual faces of junctures, because the stonemason was required to create faces that would be congruent to those of the surrounding blocks. The oldest, but also the most refined, variant was the curvilinear, also known as the Lesbian, which appeared in the seventh century B.C. Examples of this style remain at the sanctuaries of Demeter at Eleusis and Apollo at Delphi, as well as at sanctuaries in Athens; regular polygonal masonry is also found in Asia Minor and on the islands of the northeastern Aegean. This technique had a short life, and by the fifth century B.C. it had been definitively abandoned in favor of the straight joint, a less costly masonry technique.

On the acropolis of Argos, later incorporated in the structure of a medieval castle, numerous stretches of fortifications were built during the fifth century B.C. in polygonal masonry erected on top of structures from the Mycenaean period. In some areas the masonry shows the presence of horizontal faces, which perhaps were used to simplify the process of cutting the elements and to create surfaces that facilitated installation of the wall.

Trapezoidal masonry and square masonry represent further variations, distinguished by just four juncture faces on each block. With trapezoidal masonry the elements of a wall retained oblique sides, obliging the stonemason to create congruent cuts between contiguous elements, within a course of stones already somewhat regular. This technique endowed the masonry with high mechanical resistance to both pressure and seismic motion. Numerous structures dated between the fifth and the third century B.C. are characterized by trapezoidal masonry, as are many of the fortresses erected in the fourth century B.C. in the Peloponnesus and in Attica, as at the well preserved site of Eleutherai.

Square masonry, first used in the sixth century B.C., was based on the use of parallelepiped blocks installed in horizontal courses. This method notably simplified construction, both the dressing of individual elements and their assembly. The perfectly squared blocks did not require the difficult preparation of oblique joints that distinguished polygonal and trapezoidal masonry, and thus it was possible to standardize building materials: the only operation necessary to carry out at the work site was the proper placement of pieces in the wall fabric.

It was possible to execute square masonry in several ways. The irregular type was used, for example, in the substructures of the Propylaeum of Ptolemy II in the Sanctuary of the Great Gods at Samothrace, where it was characterized by the use of blocks of different heights and lengths. Two variations of this masonry type were developed that allowed the creation of uniform walls. Vitruvius (2.8.5–6) describes them as follows: "These [walls] are built in two kinds. Of these one is called *isodomum*, the other is called *pseudisodomum*. It is called *isodomum* when all the courses are built of an equal thickness; *pseudisodomum* when the courses are unequal and unlike." The isodomus wall, made with blocks of equal size, was certainly the Greeks' highest attainment in the building arts. Already present in the Archaic period, its refined alignment of junctures in alternating courses was adopted in all principal structures beginning in the fifth century B.C. The pseudoisodomic wall had courses of different heights, often alternating.

Alongside these building techniques based on the use of large stone blocks, the Greeks adopted methods based on naturally broken stone (pebbles and flat chips) or stone cut into smaller elements. These systems are known collectively as "minor work" and were widely used in the construction of private buildings and other architecture on a small scale, as well as in areas where the availability of materials made it most suitable. It was not uncommon for a mixture of techniques—both blocks and smaller elements—to be used, as demonstrated by structures with frameworks or checkered patterns.

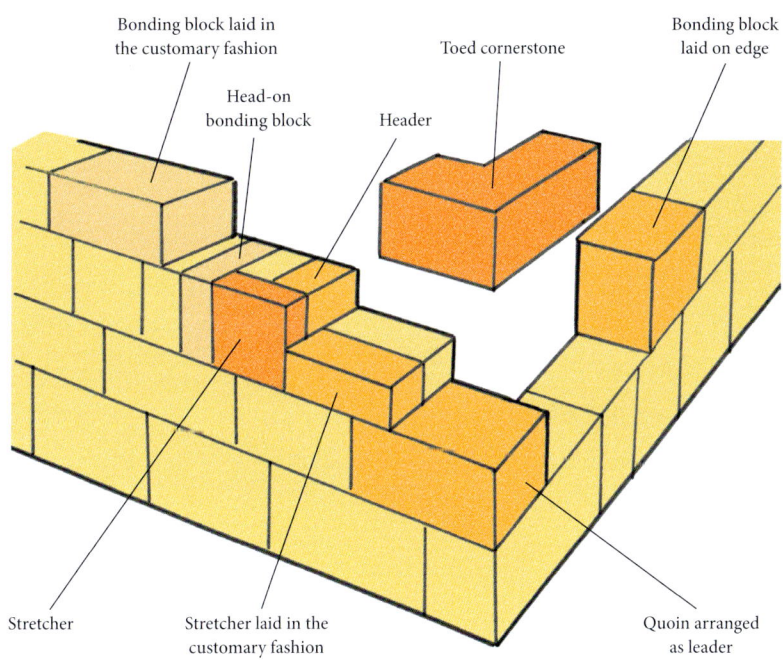

Each stone block assumed a different name according to its position in the masonry structure.

THE STRUCTURE OF WALLS IN SQUARED STONE

In his discussion of the building technique using squared blocks, Vitruvius (2.8.7) lauded Greek technical skills: "The Greeks . . . lay the stones level and put the headers and stretchers alternately. Thus they have not to fill in the middle, but with their through facing stones they render solid the unbroken and single thickness of the walls. In addition to the rest, they insert special stones facing on either front of unbroken thickness. These they call *diatonos* ['through-stones'], and they, by bonding, especially strengthen the solidity of the wall." This passage confirms the existence of a well-established construction practice, one associated with a variety of masonry types.

In many instances the Greeks made use of a double-facing wall system. This called for the interior space to be filled with a core (Gr. *emplekton*) composed of argillaceous earth and stones of various forms and sizes (including the remains from working stone blocks for the main wall). In the case of fortifications,

this filling could be more than 3 meters thick, as indicated by some stretches of the walls of Corinth and of Gortys in Arcadia.

Retaining walls (Gr. *analemmata*) presented special technical problems because of the need to offset the thrust of the embankment of a level terrace, for example, or the thrust of the terminal sections of a theater's cavea. These cases required a single facing wall, most often built with large blocks, and sometimes coupled with another internal masonry structure designed for reinforcement; this second structure remained hidden by the outer wall and thus was usually made with stone of lesser quality. Only in the second century B.C., in particular with the development of Pergamene architecture, do we find widespread use of terrace walls reinforced by the insertion of buttresses, often connected by upper arches.

As we have seen, Vitruvius emphasized the importance of the blocks of diatons that crossed the entire width of a wall. Generally, in walls of an ordinary thickness diatons bonded the two faces of the structure, both where they served as an internal filling and where courses of blocks were arranged in a double row. The use of diatons was in fact associated with the use of blocks that were narrower than the wall. In a group arranged to externally display one of the two greater faces, these blocks were called orthostats (Gr. *orthostatai*). Other stones were given conventional names based on their position inside the masonry fabric: thus there were stretchers, set so that the length of the block formed the external face, and headers, where the short end of the block appeared as the face. Greek builders developed numerous methods for assembling these elements, including isodomic or pseudoisodomic walls in squared stone made with diatons alone, walls with alternating diatons and pairs of orthostats, and structures that used blocks arranged as headers or stretchers.

When working with squared stones masons were required to pay special attention

The Erechtheum's elegant wall in isodomic opus quadratum, *constructed in blocks of Pentelic marble.*

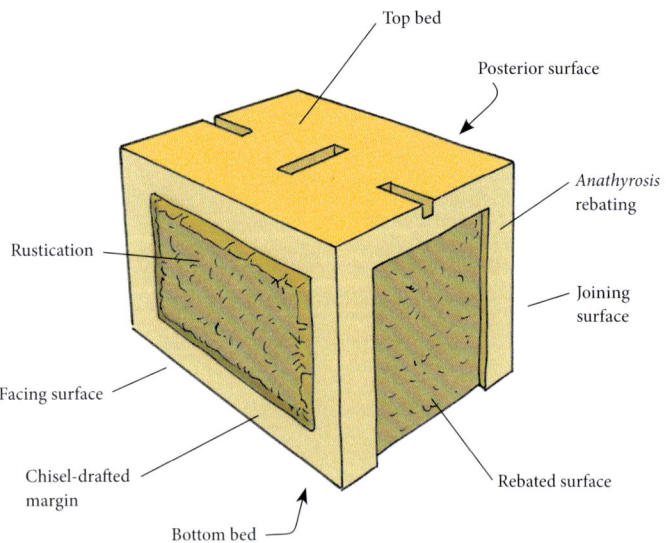

CONSTRUCTION OF A MONUMENT AND ACTIVITIES AT THE FOOT OF A WORK

As noted in chapter 1, when architectural elements arrived at the work site they had already undergone a preliminary stage of rough shaping, carried out at the quarry, to remove excess material and smooth any irregularities that resulted from extraction. A further dressing (Gr. *ergasia*) was performed at the foot of a building during its construction.

Stonecutters prepared each piece in the manner appropriate to its position within the structure, and in accord with the technical methods to be used for placing it within a building. The first operation involved elimination of the thick protective sheath (Gr. *apergon*) that had been left on the block by quarry stonecutters to preserve the block's final surfaces during its transfer to the work site. Not all of the sheath was removed, however; part of it, although reduced, was left on the faces that were destined to remain in view, to protect those surfaces during construction. In most cases only the corners—the most difficult parts to dress after the stone was mounted, because of the danger of creating irreparable chips—were given their final surface at this initial stage, together with a mark on the reference surface (Gr. *periteneia*) that indicated the eventual location of the block and that also served as a guide in the finishing stage. The faces of the block were given a definitive dressing, including preparation of the bottom and top beds as well as delineation of the type of *anathyrosis* (whether along the perimeter, on two or three contiguous sides, in a single strip, or in two parallel strips) that was appropriate to the means of joining the surfaces.

Each architectural element was then prepared to be lifted to its final position by special hoisting machines (Gr. *geranoi*). With the exception of architraves—which could be raised by means of ropes (Gr. *kaloi*) due to the open spaces between columns—all other elements of construction were intended to be set in continuous rows, one on top of the other. It

Above. The faces of a squared block.

Opposite, top. The lowest step of the crepidoma of the Temple of Nemesis at Rhamnus (430–420 B.C.) in Attica is local dark limestone, like the euthynteria, while the two upper steps are white marble from the quarries at Haghia Marina. Clearly visible on the blocks is the relationship between the protective surfaces and the lower reference bands.

Opposite, bottom. The relief of the crepidoma of the Great Temple in the Sanctuary of Apollo on Delos shows the thicknesses of protective surfaces on the blocks, still present because the building was never completed.

to dressing the blocks, since each of the six faces of a block required a different preparation, depending on how the face would be used within the masonry fabric. The facing surface (Gr. *metopon*)—the plane of a block destined to remain visible on the wall surface—was always given the most careful dressing. Thus the diatons, which crossed the entire width of a structure, presented two opposite surface faces, one on each side of a wall; the orthostats, in contrast, had a single surface face, with the opposite, posterior face hidden inside the wall.

The other faces of a block were designed for contact with adjacent blocks. The two verticals were called the contact faces (Gr. *armoi*). In most cases the surface of juncture did not extend to the entire area of the contact face but only to a narrow strip (Gr. *anathyrosis*), which was carefully smoothed by the stonecutter while the rest was summarily rebated. The lower face of a block was called the bottom bed (Gr. *basis* or *ephedra*), with reference to its role as the section of the block that rested on the course beneath it. Likewise, the upper face was called the top bed (Gr. *edra*); it met the bottom bed of the block on the course above it.

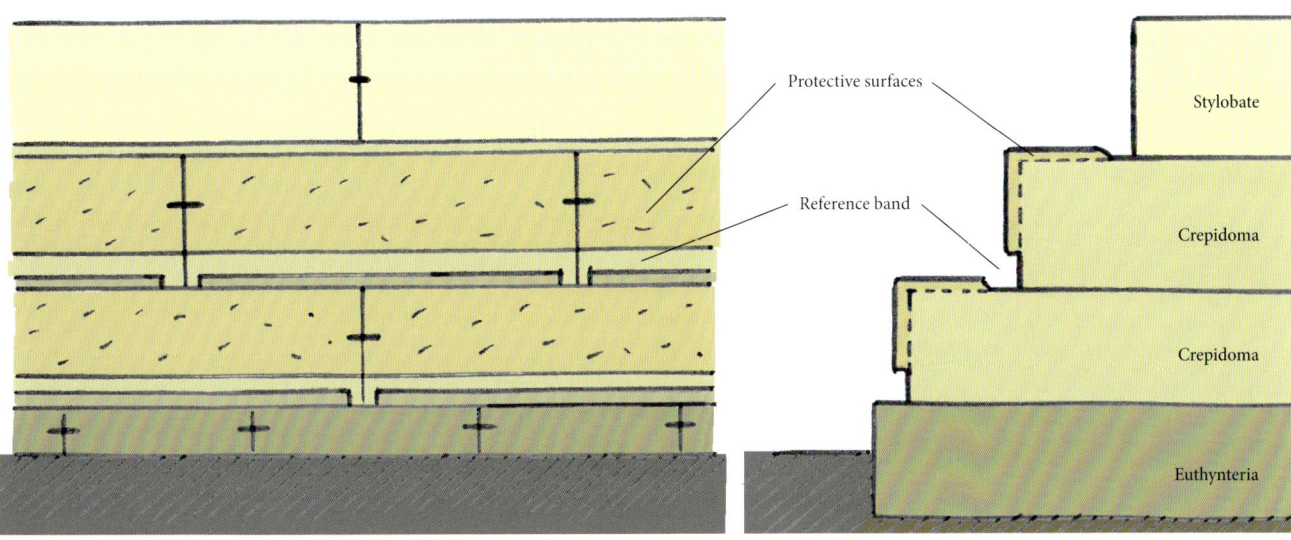

CONSTRUCTION TECHNIQUES IN THE GREEK WORLD 103

There were various ways of raising blocks to the appropriate position: A. Lifting with a cable sling (Temple of Poseidon at Sounion); B. Bosses (Propylaea of the Acropolis in Athens); C. Bosses (pre-Periclean Parthenon); D. Internal openings (Temple of Hera at Olympia); E. Internal openings (Temple of Athena Aphaia at Aegina); F. Grooves (Selinunte); G. External U-shaped canals (Temple of Athena Aphaia at Aegina); H. Lewises with grooves on the top bed (Temple of Poseidon at Cape Sounion); I. Forceps with grooves on the joining surfaces (Temple of Poseidon at Cape Sounio); J. Lewises (Temple of Hephaestus and Athena on the Agoraios Kolonos in Athens).

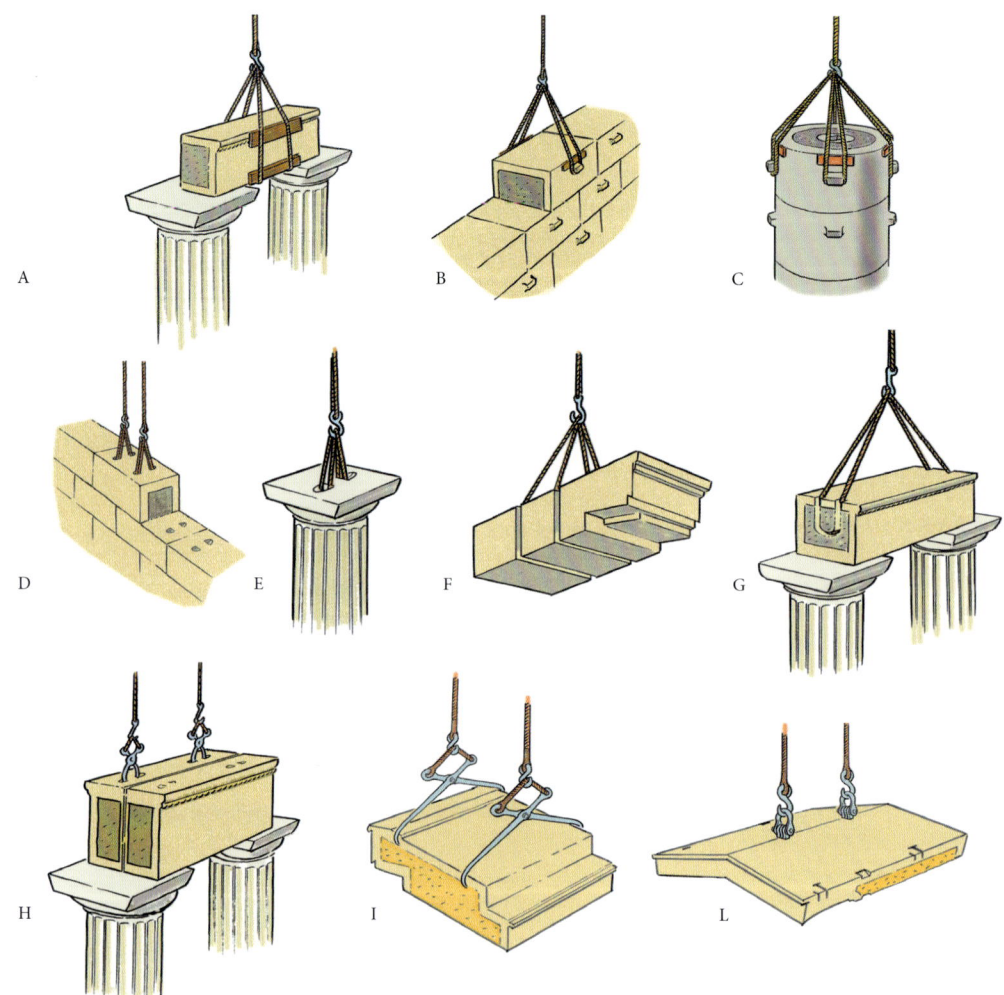

was therefore impossible to remove any ropes that passed along the bottom bed of a block after it had been installed. To deal with this problem, the Greeks developed a number of solutions, which can be divided in two main groups. The first type of solution made use of small projections that could later be removed along with the protective surfaces; the second type called for canals and grooves of various shapes to be made on the faces of a piece, and that would not be visible after completion of the work.

In the first system for lifting, blocks were raised by passing ropes around special bosses, projections for lifting that were located on the outer faces. These bosses were then eliminated during the last stage of building, when stonecutters gave the visual surfaces their final dressing. Because many work sites failed to reach that concluding stage, these projections, along with protective surfaces, can be identified on blocks that remain. Examples can be found at the construction site of the Propylaea on the Acropolis in Athens, where work was halted in 433–432 B.C. because of the Peloponnesian War, and at the temple at Segesta, which was left unfinished at the end of the fifth century, probably because of the Carthaginians' occupation of the city.

The second method for lifting was more complex, reflecting the Greeks' various ways of using canals and grooves on blocks. One of

the oldest techniques called for passing ropes through internal ring-shaped canals that had been carved in the top bed of a block. These canals characteristically took the shape of the arc of a circle or the shape of a V. Aside from the difficulty of carving the rings on the surface of the block, this method presented the drawback of subjecting a restricted area of the block to a high level of traction.

Equally ancient was the system of passing ropes through U-shaped grooves in the opposite contact faces of blocks. This method seems to have been abandoned somewhat early on, with the development of techniques that could be executed more rapidly. Similar in certain respects was the use of sling grooves, vertical canals dug on the contact faces and bottom bed of a block to permit the passage of ropes.

The sudden disappearance of methods of lifting based on the use of canals corresponds to the widespread adoption of new systems, beginning at large work sites in the fifth century B.C., that employed two different metal instruments: the lewis (Gr. *lykos*) and the forceps (Gr. *karkinos*). The first system was based on the assembly of two or three small metal parts (forming the lewis) inside a dovetailed mortise that had been made in the top bed of a block; the parts assumed a shape congruent with that of the groove, and were then joined by a metal pin that made it possible to connect a rope for lifting.

In the Doric temple at Segesta (430–420 B.C.), lifting bosses on the blocks of the crepidoma remained in place because of an interruption at the work site, perhaps as a result of the city's capture by the Carthaginians.

CONSTRUCTION TECHNIQUES IN THE GREEK WORLD

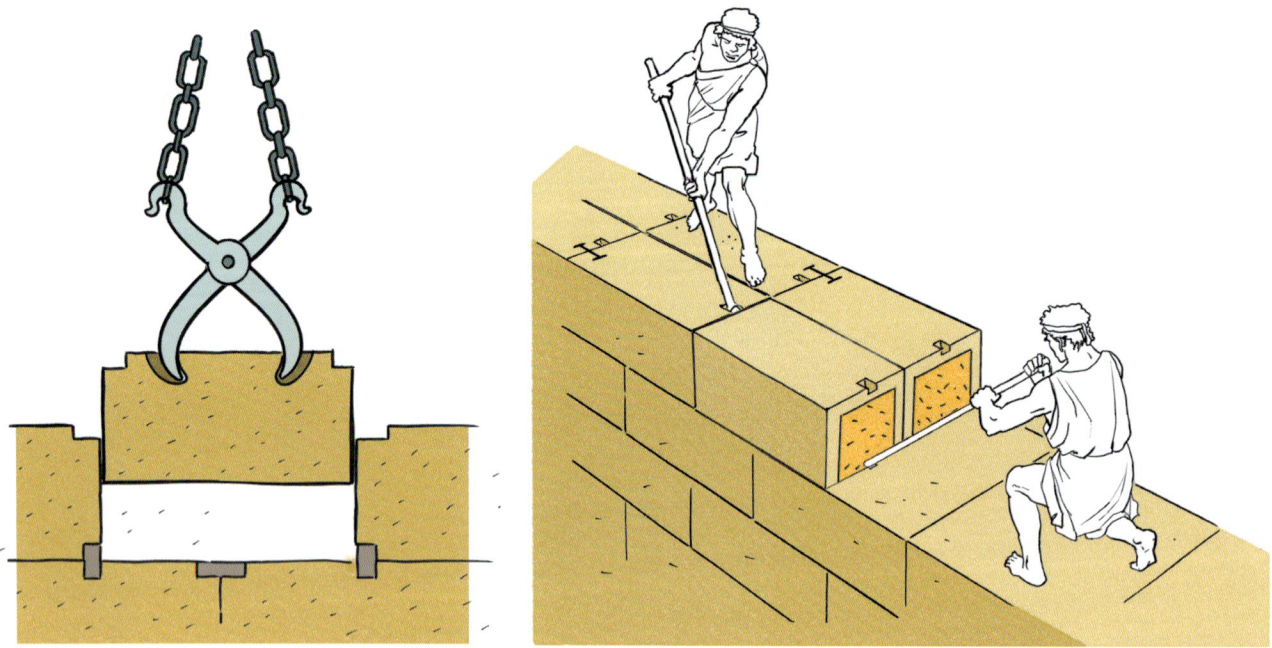

Left. Reconstruction of the use of forceps based on holes found on the top bed of several blocks of the Parthenon (447–438 B.C.).

Right. The use of metal levers for final positioning of blocks in a wall of isodomic opus quadratum.

The second method, widely used in Roman as well as Greek architecture, was the forceps, mentioned by Vitruvius (10.2.2) in relation to lifting machines (Lat. *machinae tracctoriae*): "To the bottom of the block, iron pincers are fixed, the teeth of which are adjusted to holes in the block of stone. Now when the rope has its end tied to the windlass, and the handspikes draw and turn the windlass, the rope in winding around the axle is made taut and so lifts up weights to their place in the work." In other words, these curving pincers were inserted into the designated holes, and when drawn by the ropes they grabbed the block and held it securely in place to allow lifting.

The stonecutter's activity did not end once the block was prepared to be raised into position. Additional grooves were required on the various faces of a piece so that it could be maneuvered into the correct position and connected to other elements of the building. After a block had been lifted and placed in the intended course, it was then carefully positioned to reach the contact faces of contiguous architectural elements. This apparently simple operation was in practice quite complex because of the great weights connected to the sizes of the pieces. At the same time the movement posed serious risks because the edges of the stone could easily chip, rendering the piece unusable. For these reasons, at least as early as the Classical period, the Greeks refined the use of metal levers (Gr. *mochloi* or *loisthroi*) and perfected a system employing corresponding grooves, usually created by the stonecutter in the bottom bed and along the edge of the top bed.

Other grooves were made in the stone to hold the lead connecting pieces that joined it to contiguous architectural elements, and that gave the entire structure greater stability. There were two types of such joining: the first was horizontal and involved the use of clamps (Gr. *desmoi*) inserted between two elements in the same course; the second was vertical and based on the insertion of pins (Gr. *gomphoi*) between two superimposed pieces. As one would expect, the clamps and pins did not have consistent shapes throughout ancient history, and their forms varied with

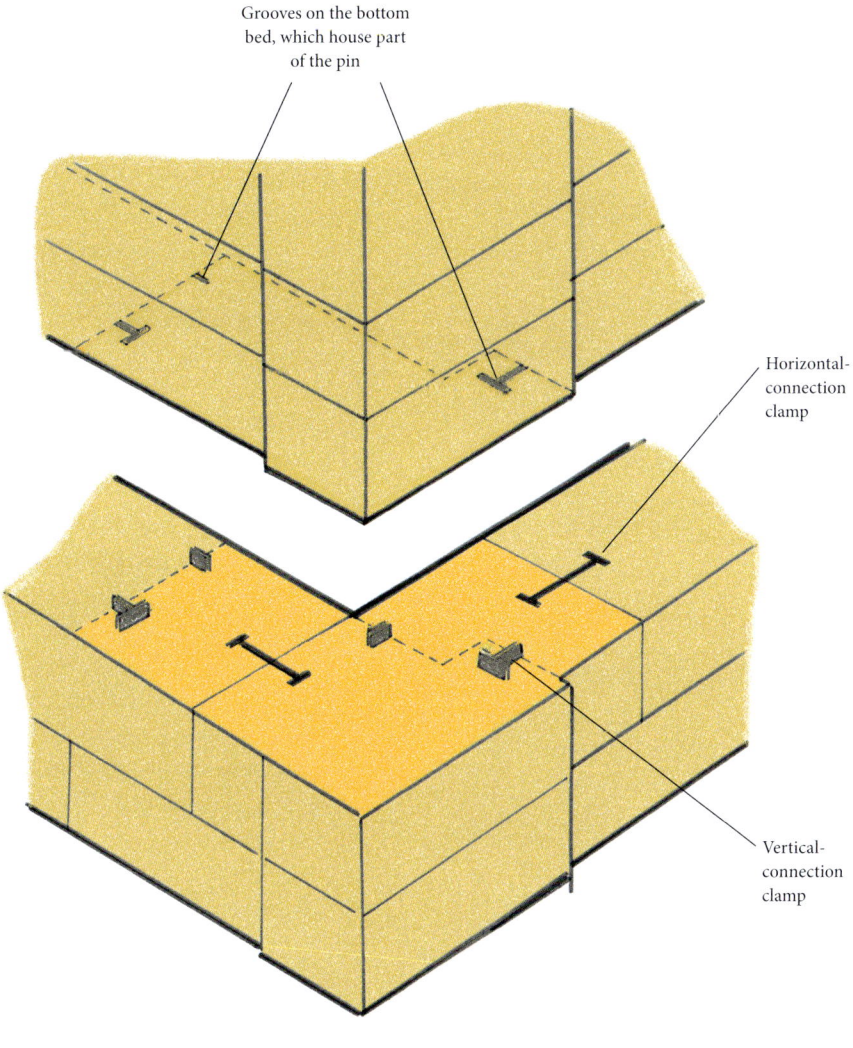

Grooves on the bottom bed, which house part of the pin

Horizontal-connection clamp

Vertical-connection clamp

Top. Position of horizontal-connection clamps and vertical-connection pins in a building on Delos made in opus quadratum.

Bottom. The main types of clamps for the longitudinal connection of blocks.

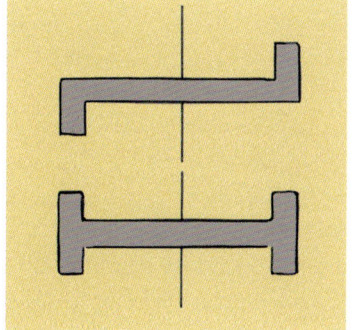
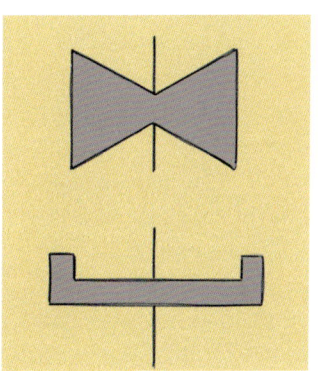
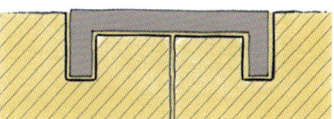

CONSTRUCTION TECHNIQUES IN THE GREEK WORLD 107

This profile of the base of a column was incised on a wall of the Temple of Apollo at Didyma to guide stonecutters in their work.

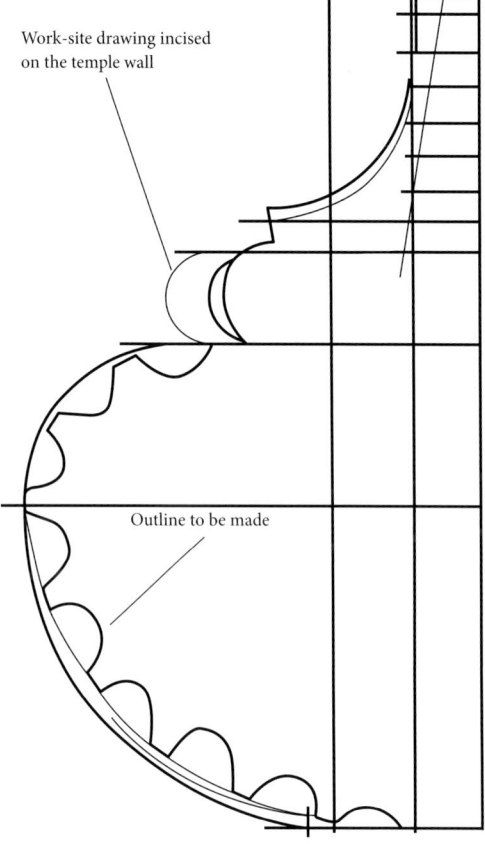

Work-site drawing incised on the temple wall

Outline to be made

the chronological period and geographical context of individual work sites.

Each clamp or pin fit into a groove with a matching shape, but of a slightly larger size. Horizontal connections were generally made along the top bed of the two elements; since the clamp was inserted between the two, the groove in each of the blocks was made to hold only half of the clamp. The stonecutter would create grooves with different shapes according to the clamp's form. The oldest types were the double gamma, double T, and dovetail (Gr. *pelekinoi*), but the hook, or Greek Pi (Gr. *peronai*), and U-shaped clamps soon came into use. In most cases these clamps were made of metal (iron or bronze) that was then sealed in the groove by pouring in a sheath of lead.

The pins used in vertical connections were also made of metal except when the connection related to two superimposed column drums, in which case the pins were often made of wood (Gr. *empolion*). Metal pins of this type were generally placed in an axial position, and required the creation of a rectangular or circular socket into which the pin was inserted. Individual pins could be shaped in a single piece, or composed of two elements that were assembled and then joined vertically to a third element (Gr. *polos*) with a quadrangular or circular section, resulting in a cone-like configuration.

ORGANIZATION OF THE WORK SITE IN THE GREEK WORLD

Plutarch (*Life of Pericles* 12.6–7) portrays a lively image of the myriad materials and workers at one of the greatest work sites of the ancient world, that which transformed the face of the Athenian Acropolis during the second half of the fifth century B.C.:

> He [Pericles] boldly suggested to the people projects for great constructions, and designs for works which would call many arts into play and involve long periods of time. . . . The materials to be used were stone, bronze, ivory, gold, ebony, and cypress wood; the arts which should elaborate and work up these materials were those of carpenter, molder, bronze-smith, stonecutter, dyer, worker in gold and ivory, painter, embroiderer, embosser, to say nothing of the forwarders and furnishers of the material, such as factors, sailors, and pilots by sea, and, by land, wagon makers, trainers of yoked beasts, and drivers. There were also rope makers, weavers, leather workers, road builders, and miners. And since each particular art, like a general with the army under his separate command, kept its own throng of unskilled and untrained laborers in compact array, to be as instrument unto player and as body unto soul in subordinate service, it came to pass that for every age, almost, and every capacity the great

One of the bases of the Ionic columns of the Temple of Apollo at Didyma, characterized by the horizontally grooved profile of the torus.

abundance was distributed and scattered abroad by such demands.

In general, each ancient work site included a great number of professional figures and artisans. The director of the project, or at least of its execution, was the architect (Gr. *architekton* or *mechanikos*), who was sometimes associated with a collaborator, the vice-architect (Gr. *hyparchitekton*). Some architects (Gr. *technitai*) possessed specialized skills in the fields of civil and military engineering, including installation of the monumental cranes used to lift architectural elements. Alongside the architect at public sites were a secretary (Gr. *grammateus*), as well as magistrates (Gr. *epistratai*) in charge of supervising the work and keeping records; in addition, there were official accountants (Gr. *logistai*) responsible for recording expenses of the construction in a special report (Gr. *syngraphe*).

These reports were engraved on slabs for a public record of the sums spent, and some have survived to provide a great deal of information on the organization of Greek work sites and the costs of construction. For example, the report relating to the Temple of Asclepius in the Sanctuary of Epidaurus, dated to the fourth century B.C., records an expense of 4,320 drachmas for extraction and transportation of stone material for the temple's access ramp and floor; 3,068 drachmas for mounting the peristyle and trabeations; and 1,336 drachmas for grooving the column shafts.

And who performed all these operations? The answer is that mass of artisans (Gr. *demiourgoi*) who formed the true soul of the work site. Among them were stonecutters (Gr. *lithologoi*) in charge of dressing stone surfaces (Gr. *epixestes*); carpenters (Gr. *tektones*); specialists in working marble (Gr. *marmaropoioi*); as well as many others.

CONSTRUCTION TECHNIQUES IN THE GREEK WORLD

This metrological relief found at Salamis indicates measurements for the foot (A, F), half-foot (E, G), span (B, H), cubit (C), and orthodoron (D, J).

Reports concerning the completion of the Erechtheum between 409 and 407 B.C. have made possible a reconstruction of the work-site population involved in the undertaking. Out of a total of 107 artisans, 24 were Athenians, 42 were foreigners (recorded with an indication of their areas of origin), and 20 were slaves. Some of these workers were paid a daily salary (Gr. *misthos*), while others were paid on a piecework basis. This was the case for the site's two groups of stonecutters, who earned 350 drachmas for making the grooves in the six columns of the building's eastern facade. Assistants and lower-level workers (Gr. *hypourgois*), such as those involved in erecting and dismantling scaffolding, were paid a half rate, while the workers earning the most money were, as one would expect, the sculptors. It has been calculated that the creators of the marble reliefs for the frieze at the Erechtheum earned as much as 3 to 5 drachmas for each day of work.

The basis for most major architectural creations was the architect's plan (Gr. *prokentema*), sometimes accompanied by a three-dimensional model (Gr. *paradeigma*), which might depict the entire building or only some of its details on a larger scale. The Erechtheum's reports record a payment of 8 drachmas for a wax model of the rosettes and acanthus leaves to be reproduced in the marble decoration. An inscription from another site, dated to the fourth century B.C., requests that a stone model be brought from Delos to Athens for the sculptors' reference when creating the capitals of an Ionic temple.

Work-site drawings have also been found on stone surfaces at various monuments, and no doubt served as models for the execution of a building's individual parts. Various sections of the great Temple of Apollo at Didyma preserve traces of such engraved designs, which date back to the middle of the third century B.C. One of these designs delineates the contour of the base for a specific column; two other designs relate to the shafts of columns (one shows the position of the grooves, the other the ratio of tapering with an indication of the entasis); a fourth engraving is a drawing of the shape of one of the ceiling coffers.

While individual details and work-site drawings illustrate the forms of a building's various parts, it was the measurement system that laid the groundwork for a project overall and for every individual construction activity. The unit of measurement was the foot (Gr. *poys*), composed in turn of multiples also related to parts of the human body: the half-foot (Gr. *hemipodion*); the palm (Gr. *palaiste*), equal to one-quarter of a foot; the digit (Gr. *daktylos*), equal to one-sixteenth of a foot; the span (Gr. *spithame*), equal to three-quarters of a foot; the cubit (Gr. *pechys*), equal to one and a half feet; and so on.

The anthropometry of this system is illustrated in several intriguing ancient metrological reliefs, such as that found at Salamis and the one today preserved at Oxford University. Over time, however, the basic unit of the Greek world varied: while the Attic-Cyladic foot measured 29.4 centimeters, the Doric was equivalent to 32.6 centimeters, and the Ionic extended to 34.8 centimeters.

CONSTRUCTION TECHNIQUES IN THE ROMAN WORLD

Following the expulsion of the last Tarquin king in 509 B.C., the Romans inaugurated the largest Tuscan temple ever made—dedicated to Jupiter, Juno, and Minerva—on the southern summit of the Capitoline Hill. The architecture employed was still that of the Etruscan world. In fact, decoration of the building—including a great terracotta quadriga, or chariot, that crowned the roof—was the work of an Etruscan artist, a certain Vulca, sculptor of Veii. Rome, however, was destined to affirm its presence and language over the entire Mediterranean region, with a history that began with the destruction of Veii itself in 396 B.C. Over the span of two centuries, Roman architecture succeeded in drawing ideas from the Greek and Magna Graecian world to express an autonomous identity that reflected Latin society and its cults. By the elaboration of specific construction techniques and continuous experimentation, in a short time Roman architects were able to create great structures that revolutionized the very concept of architectural space throughout the ancient world. From temples to basilicas, and from imperial residences to bathing complexes, theaters, and amphitheaters, every structure spoke a "Roman" language. Its lexicon was soon adopted by all the cities of the empire in a continuous process of emulation that transformed those urban spaces, in the words of Aulus Gellius (*Attic Nights* 16.13.9), into "miniatures, as it were, and in a way copies" of Rome.

CONSTRUCTION WITH LARGE BLOCKS

Just as in Greece, the oldest monumental structures in the Italic world were massive walls built in defense of cities. These structures were made without the use of lime mortar, and the large blocks of stone, which in the earliest periods were given only the most rudimentary shaping, were set directly on top of each other, as in the great Cyclopean fortifications found at Amiternum, birthplace of the historian Sallust, and at Arpinum, birthplace of the philosopher and statesman Cicero. In these structures the irregular shapes of individual blocks, set in place without dressing, resulted

A stretch of the polygonal-masonry city walls at ancient Cosa, near Ansedonia, in Tuscany.

Inside the Nuovo Museo Capitolino on the Capitoline Hill are preserved remains of the foundations of the temple dedicated in the sixth century B.C. to the triad Jupiter, Juno, and Minerva. Built in blocks of cappellaccio, the structure reached a height of about 5 meters.

in discontinuous junctures and interstices that required infilling with smaller stones.

The Romans and other Italic peoples soon began to cut stones in order to create structures with a more organic sense of form. Thus was born the building technique known as polygonal masonry, the several types of which are identified according to the shape of the blocks used. Irregular polygonal work was characterized by stone that continued to be hewn in a rough manner, and as a result junctures were still imprecise and construction of regular horizontal courses was not possible. Some scholars have argued that in these cases the outer faces of blocks were hewn only after the blocks had been set in place. This technique, used in the walls of Praeneste (modern Palestrina) and of Signia (Segni), was also adopted in nondefensive structures, such as the walls of the Temple at Bovianum Vetus (Pietrabbondante) and the terraces along the Via Appia between Anxur Tarracina (Terracina) and Fundi (Fondi).

Regular polygonal work required greater care with individual junctures, since it was necessary to create congruent faces between adjacent blocks. Stone elements thus assumed the shape of regular polygons (trapezoidal masonry), sometimes with concave or convex sides; at the corners, however, joints tended to assume a more substantial, rectilinear form. Examples can be found in the walls of Cosa (modern Ansedonia) and in some terraces of the Sanctuary of Fortuna Primigenia at Praeneste (modern Palestrina).

The most commonly employed construction technique using large blocks was *opus quadratum*, squared blocks laid in parallel courses without mortar. This method appeared in Rome as early as the sixth century B.C. in the foundations of the Temple of Jupiter Capitolinus, on the Capitoline. These structures, preserved at a height of about 5 meters, were built with blocks of cappellaccio quarried from the slopes of the hill. The first city walls used the same technique and material, and remains are still visible below Santa Sabina and near the Baths of Diocletian. After the sack of the city by the Gauls in 390 B.C., when the Romans built the powerful fortifications known as the Servian Wall in blocks of tufa from the Grotta Oscura, they again adopted the technique of "rectangular stones" (Lat. *saxo quadrato*) mentioned by Livy (*History of Rome* 6.32.1) in reference to contracts for the stones in 378 B.C. The new walls, up to 10 meters high and about 11 kilometers long, were erected in large blocks about 59 centimeters on each side, equal to two Roman feet (1 pes equals 29.6 centimeters), arranged as headers and stretchers (on edge). Various crews worked on the wall simultaneously, as indicated by irregularities at the meeting points between stretches. Many blocks bear incised symbols and letters that are quarry or measurement marks related to work-site operations.

The method of making Roman walls in *opus quadratum* did not differ enormously from that we have seen used in Greek architecture. There were diatons and orthostats, blocks arranged on edge or as headers, and blocks installed in identical or alternating courses.

Between the second century B.C. and the imperial period, *opus quadratum* was adopted for the construction of numerous buildings, sometimes in an autonomous form, without the blocks acting as a facing wall for a masonry core. For example, the long portico erected at the foot of the Capitoline, along the stretch of the triumphal route that crossed the Forum Holitorium, was built entirely in blocks of peperino and travertine. The structures in *opus quadratum* at the imperial forums appear far more monumental to us today, as in the Forum of Caesar, where behind the southwestern colonnade of the square, a series of shops (Lat. *tabernae*) on several levels were built mostly with blocks of tufa and travertine, materials also used for the great radial-block arches. *Opus quadratum* was also used to create the large pillars behind the Temple of Venus Genetrix, erected by pairing the two elements in alternating courses. Equally impressive are the remains in *opus quadratum* at the Forum of Augustus, beginning with the wall in blocks of peperino, pietra gabino, and travertine that, at the height of about 30 meters, closed the complex on its rear side, facing Subura. In addition, the outer walls at the nearby Forum of Nerva (also known as the Forum Transitorium) were made in squared blocks of lithoid tufa, with the columns, trabeation, and attic in Luna marble.

In the Forum of Caesar, completed in the Augustan period, a series of shops were built on several levels, using blocks of tufa and travertine with large arches in radial blocks.

CONSTRUCTION TECHNIQUES IN THE ROMAN WORLD 113

At Ephesus in modern-day Turkey, the great gate of the Tetragonos Agora was erected in the Augustan period following the model of a monumental triumphal arch, using square blocks of white marble and cornices in gray-blue stone.

Opposite. The monumental rear wall of the Forum of Augustus, about 30 meters high.

Innumerable monuments in *opus quadratum* can be found throughout the empire. From Spain to North Africa, from Greece to the eastern provinces, many cities adopted this technique, often in the wake of ancient local building traditions. At Ephesus the great portal of the Tetragonos Agora was erected in the Augustan period on the model of a triumphal arch with three arches, using square blocks of white marble; only in the arcades were moldings in gray-blue stone inserted to create an elegant polychrome effect.

ROMAN *OPUS CAEMENTICIUM*: A TECHNICAL REVOLUTION

As we have seen, the Greek building tradition included the custom of making walls composed of two facing walls filled by a core (Gr. *emplekton*) of argillaceous soil and stones. The Romans soon intuited that by replacing the argillaceous soil of the core with a good-quality lime mortar, they could build a wall of greater strength. Thus they developed *opus caementicium*, to which Vitruvius (5.12.5) refers: "The work is to be filled in with concrete of stone, lime, and sand [Lat. *structura ex caementis calce et harena*]." Unlike the soil used by the Greeks as a binder, during the phase of crystallization lime transformed the mortar into "artificial stone," firmly fixing not only the mixture's fine aggregates (sand, pozzolana, and ground bricks) but also the larger stones inserted in the walls.

Opus caementicium was a filling material, usually made to form the core between two facing walls. To build walls in *opus caementicium* Roman masons first made two external rock faces; treating these walls like a permanent mold, they then filled the interior with an *opus caementicium* composed of various kinds of aggregates bound by mortar. In his description of the technique Vitruvius (2.8.7) continued to use the Greek term *emplekton* (in a variant spelling): "The second [method] is that which

Brick facing walls of the residence of the Flavian emperors on the Palatine Hill. The internal core in opus caementicium *is visible, as is the careful arrangement of the stones inside it.*

they call *enplecton*, which our country people still use. In this the faces are dressed; the rest of the stones are laid with mortar in their natural state, and they bond them with alternating joints. But people nowadays, being eager for speedy building, attend only to the facing, setting the stones on end, and fill it up in the middle with broken rubble and mortar. Thus three slices are raised in this walling, two of the facings, and a middle one of the filling in."

The aggregates (Lat. *caementa*) used in *opus caementicium* were of diverse kinds, from pebbles to irregular bits of stone, from brick fragments to other terracotta goods such as broken pieces of amphorae. Like the mortar with which it was mixed, these materials were far more economical than squared stone, and their use in *opus caementicium* provided walls with significant static capacities. Due to these qualities, as well as the particular versatility of *opus caementicium* at the work site, the technique became Rome's most typical way of building and was exported throughout the empire—constituting a true construction revolution, handed on to the Middle Ages and continuing in use until modern times.

The technique of *opus caementicium* began spreading from Latium and Campania around the end of the third century B.C. By the first half of the following century, Cato the Elder (*On Agriculture* 14.4–5), writing about the construction of rural houses, prescribed its use for building foundations: "In a steading of stone and mortar groundwork, carry the foundation one foot above ground, the rest of the walls of brick; add the necessary lintels and trimmings." Some houses at Pompeii from the Samnite period—such as the House of the Surgeon (dating to the end of the fourth century B.C.) or that of Sallust (third century B.C.)—have walls made with a filling based on mortar, although one that is still very earthy and full of lumps of lime. *Opus caementicium* seems to have been introduced to Rome itself somewhat later. Between the end of the third and the beginning of the second century B.C., the technique was used on the Palatine to build the structure of the temple dedicated to Magna Mater, the Great Mother Cybele. As a result of recent studies, there is discussion today on its use in the great Porticus Aemilia, the monumental structure built in 193 B.C. near the port on the Tiber by the two aediles L. Aemilius Lepidus and L. Aemilius Paullus. New interpretations of the remains discovered in the nineteenth century to the south of the Via Marmorata, and a variant reading of their depiction on the *Forma Urbis Romae* (the great marble map from the Severian period), now make it impossible to consider this building as the oldest monumental example of the use of *opus caementicium*.

A long process of experimentation was necessary to perfect the execution of *opus caementicium*. This involved not only exploration of its applications but also trials of the types of aggregates used, the ways in which they should be handled, and the proportions of the mortar's components. In this regard, Vitruvius (2.8.2–3) wrote that walls should "be built with very minute stones; so that the walls, thoroughly saturated with mortar of lime and sand, may hold longer together. For since the stones are of a soft and open nature, they dry up the moisture by sucking it out of the mortar. But when the supply of lime and sand is abundant, the wall having more moisture will not quickly become perishable, but holds together. When once, also, the moist power has been sucked out of the mortar, through the loose structure of the rubble, and the lime separates from the sand and is dissolved, the rubble also cannot cohere with them, but renders the walls ruinous with lapse of time." He drew attention to "some tombs that are built near the city, faced with marble or squared stone, and, in the interior, constructed with walling material pressed down." In these cases, Vitruvius noted, "the mortar becomes perishable in time and is drawn out through the loose joints of the rubble. Hence the tombs collapse and disappear when the union of the joints is broken by settlement."

Depending upon the thickness or the static function of a wall inside a building, *opus caementicium* could be made in any of several ways. The simplest technique, generally adopted for walls with an ordinary thickness (from 30 to 75 centimeters), called for pouring a layer of mortar into the nucleus of the wall, setting the caementa within it, and then covering them with another layer of mortar. The result was a homogeneous structure of good quality, characterized by the horizontal arrangement of the internal caementa. In other cases, such as the filling of a larger space, the alternation of mortar and aggregates was more straightforward, since the mason added the caementa to an abundant layer of mortar rich in water, ramming it down to eliminate any possible air pockets. A third system, far more demanding and less widespread, called for premixing the *opus caementicium* on site and then inserting it between the facing walls.

FOUNDATIONS IN *OPUS CAEMENTICIUM*

As we have discussed with regard to Greek architecture, the creation of any building began with the construction of its foundations (Lat. *fundamenta*), and their importance

The structure of one of the pylons of the aqueduct that served the baths in Domitian's palace on the Palatine. The masonry was strengthened by using exclusively brick fragments in the concrete core.

The strong foundations in opus caementicium of the Temple of Venus in Rome, built by Hadrian during the first half of the second century A.D.

solid and in the solid, as may seem proportionate to the amplitude of the work, of a breadth greater than that of the walls which shall be above the ground; and these foundations are to be filled with as solid a structure as possible."

According to Vitruvius's prescriptions, in order to make solid foundations it was necessary to properly prepare the terrain where the building would stand, beginning with an excavation that reached bedrock or at least a layer of clay with good resistance. This rule was not always followed by builders, and the results could prove catastrophic. The historian Tacitus (*Annals* 4.62–63) tells the story of what happened in A.D. 27 at an amphitheater at Fidenae, near Rome: "A certain Atilius, of the freedman class, who had begun an amphitheater at Fidena, in order to give a gladiatorial show, failed both to lay the foundation in solid ground and to secure the fastenings of the wooden structure above; the reason being that he had embarked on the enterprise, not from a superabundance of wealth nor to court the favors of his townsmen, but with an eye to sordid gain. The amateurs [fans] of such amusements, debarred from their pleasures under the reign of Tiberius, poured to the place, men and women, old and young. . . . This increased the gravity of the catastrophe, as the unwieldy fabric was packed when it collapsed, breaking inward or sagging outward, and precipitating and burying a vast crowd of human beings, intent on the spectacle or standing around. . . . Fifty thousand persons were maimed or crushed to death in the disaster; and for the future it was provided by a decree of the senate . . . that no amphitheater was to be built except on ground of tried solidity." In other cases, the poor state of foundations was evident even during construction of the building. Early in the second century A.D., when Pliny the Younger was governor of Bithynia, he wrote a letter to the emperor Trajan (*Epistles* 10.39) that expressed his concerns regarding the instability of a building in his territory: "The theater at Nicea, sir, is more than half

to the endurance (Lat. *firmitas*) of the entire structure was well known to the Romans. In terms of a structure's successful outcome, foundations were equal in value to the quality of the construction materials used, and it was necessary that they were made deep enough in the ground to reach a compact and firm (Lat. *solidum*) layer, as Vitruvius (1.3.2) advised: "Account will be taken of strength when the foundations are carried down to the solid ground, and when from each material there is a choice of supplies without parsimony." With respect to building the walls and towers of a city, he added (1.5.1), "The foundations of the towers and walls . . . are to be dug down to the

Method of making foundations in opus caementicium *inside a reinforced trench.*

built but is still unfinished, and has already cost more than ten million sesterces, or so I am told—I have not yet examined the relevant accounts. I am afraid it may be money wasted. The building is sinking and showing immense cracks, either because the soil is damp and soft or the stone used was poor and friable. We shall certainly have to consider whether it is to be finished or abandoned, or even demolished, as the foundations and substructure intended to hold up the building may have cost a lot but look none too solid to me."

Making foundations in *opus caementicium* followed very different methods from those involved in the use of squared blocks. As we have seen in terms of Greek architecture—and the technique was also adopted by the Romans before the extensive use of concrete—when building foundations in stone blocks, it was first necessary to dig a trench wider than the foundations themselves to permit the movement of workers at the site. For foundations in *opus caementicium* the semifluid nature of the material made it possible to dig a trench (Lat. *fossa*) equal to the size of the structure being built. Indeed, it was the open hole itself that acted as a mold for pouring the concrete. These open-trench foundations were common whenever the terrain was sufficiently compact and the trench did not require much depth. In other situations it was necessary to reinforce the walls of the trench to prevent them from collapsing during digging or in the period before the concrete was poured. The walls were therefore shored up using horizontal wooden planks, or shutters, that were supported by a framework of vertical posts and crosswise struts. Having made this shuttering and rendered the trench stable, builders could continue digging to reach, with further shoring, the necessary depth. The wooden structure thus assumed the nature of a formwork that would not be reused, and impressions of its wooden elements would be left in the concrete long after they had rotted away. Such foundations, known as reinforced-trench foundations, on occasion reached monumental dimensions. Still impressive examples include the Temple of Venus and Roma in Rome.

In cases of sloping terrain or when the floor level would be higher than the surrounding land, a segment of the foundations would have

Below. Cross section of the Flavian amphitheater showing the system of substructures used. The colors identify the different construction materials, and the numbers refer to the positions of the travertine uprights.

Opposite. The Flavian Amphitheater, better known as the Colosseum, was the largest amphitheater in the Roman world. Its construction was begun by Vespasian in A.D. 70 on the site of the artificial pond of Nero's Domus Aurea.

to be raised higher than the level reached by the trench. This was accomplished by extending them upward with normal facing walls—true walls made like those of the elevation of the building but destined to be buried. Formworks were commonly used for this section of the foundations, as indicated by impressions left in the mortar by the wooden pieces. These formworks differed from shuttering used in trenches, primarily in terms of the position of the support struts with respect to the frame: in the trench the struts served to hold back the thrust of the trench walls, remaining there to be buried in the concrete; in the elevation they served instead to contain the thrusts of the material itself and were thus located in an exterior position.

FOUNDATIONS OF THE FLAVIAN AMPHITHEATER

In Roman and Greek architecture, the foundations of a structure were of two types, discontinuous and linear. In the first, common to the construction of aqueducts on arches, foundations were reduced to a succession of individual pylons (Lat. *pilae*); in the second, foundation walls were created to receive the load of elevation walls so that they exactly reproduced the planimetric outline of the building. In some cases, such as the colonnades of temples, these walls were reinforced at the points where the structure's load would be concentrated by means of a different kind of wall or by the use of stone blocks. Such arrangements are called linear foundations with links.

More complex are trellis foundations, built when the unstable nature of the terrain required consolidation of the lower area. In such cases foundation walls were made that crossed to form caissons, which were then packed with backfill material. These masonry cells often were connected or covered by means of a series of arches or vaults. This system was widely used in substructures and large podiums (high masonry bases) in such a way that the thrust of the backfill was divided and distributed among the internal walls.

Without a doubt, the most impressive foundations were those based on the construction of a concrete bed (Lat. *solum*). These consisted of a compact layer in *opus caementicium* that extended for most if not the full extent of the construction. One of the most interesting applications of this system is seen in the gigantic, articulated foundations of the Colosseum, the amphitheater erected by the Flavian emperors in the area previously occupied by the artificial pond of Nero's Domus Aurea. Work on the complex began during the reign of Vespasian in A.D. 70 and was completed a decade later by the emperor Titus with the exception of the decoration, which required several additional years and was only finalized during the reign of Domitian.

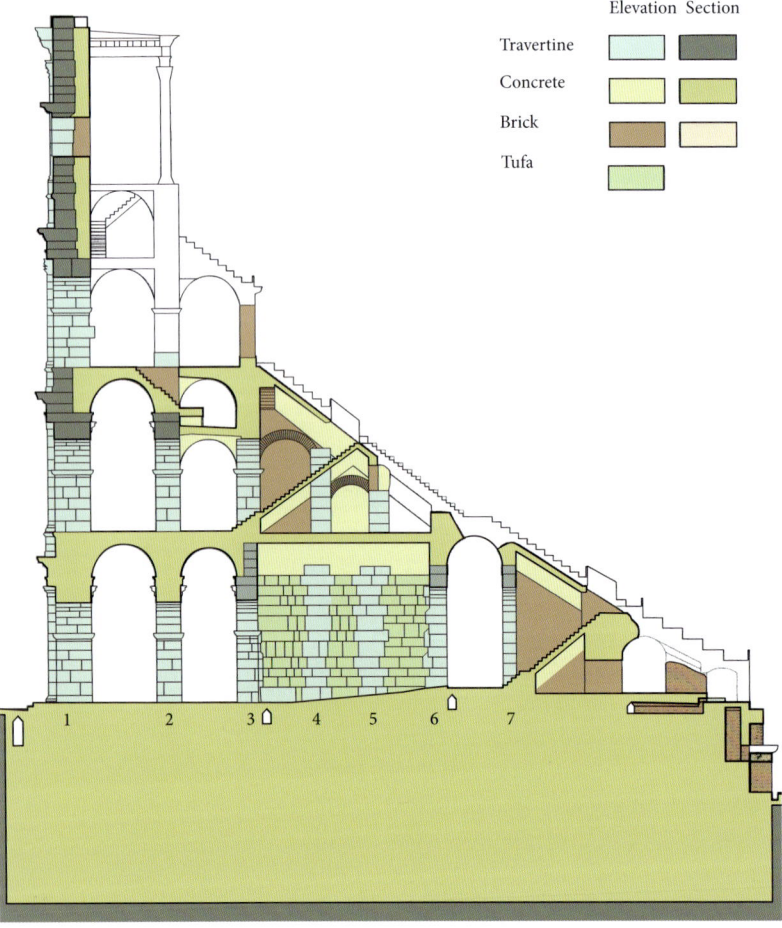

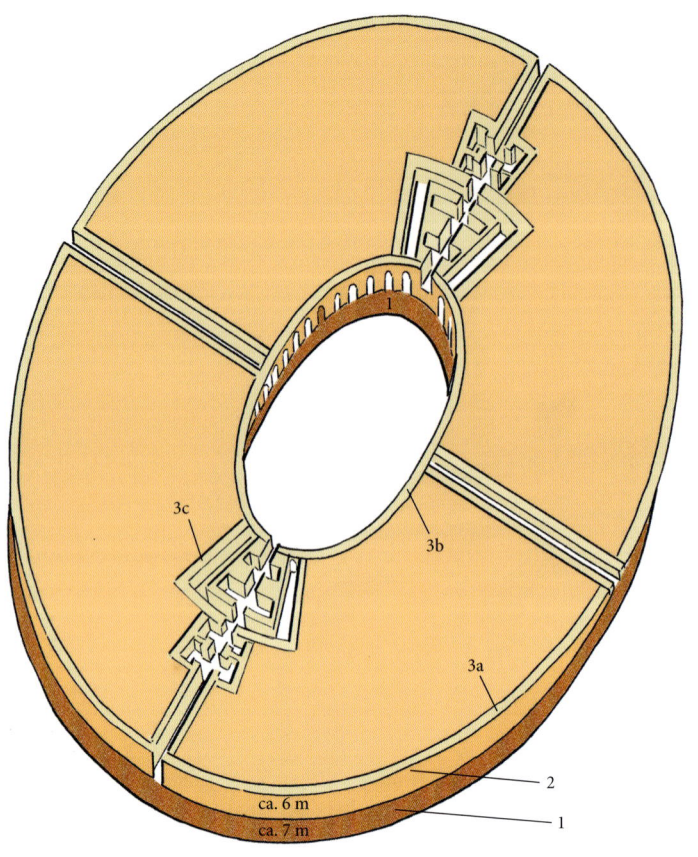

Diagram of the foundation of the Colosseum: 1. Lower ring; 2. Upper ring; 3a–c. Containment walls of the upper ring.

The Colosseum, a monumental structure with an elliptical shape, is the largest amphitheater in the Roman world. Its main axis measured about 188 meters, while its minor axis reached 156 meters, and its height, including the three architectural orders of external arches as well as the crowning attic, extended a full 50 meters. Construction of the monument was part of the massive building program undertaken by the Flavian dynasty—which sought to return to the city of Rome the spaces that had been usurped by Nero for his private use—and involved an immense work site and unprecedented quantities of construction materials: more than 100,000 cubic meters of travertine and about 300 tons of iron were required just for the clamps used to connect the blocks; the quantities of bricks and mortar ingredients employed are incalculable.

To build such a monument, its architect had to prove himself a true master of engineering. As noted, the Colosseum stood on the site of the giant artificial pond of Nero's Domus Aurea, which held more than 34,000 square meters of water at a depth that reached three or four meters. Not surprisingly, the terrain was highly swampy and required enormous efforts at consolidation. Since the gigantic concrete-bed foundation did not extend beneath the area of the arena itself, it assumed the shape of the above-ground structure, that is, a large elliptical ring. The dimensions of the foundation were monumental: it formed a strip 60 meters wide and 13 to 14 meters deep on an ellipse of 196 by 164 meters, for a total surface area (excluding the arena) of 22,665 square meters. Digging the hole for these foundations alone required the removal of roughly 269,000 cubic meters of earth, an operation that called for at least 3,000 workers who were employed for more than a year, not counting those involved in digging the earth and those who built the hole's reinforcement.

The foundations were made of two superimposed elements. The lower ring, about 7 meters deep, was built in *opus caementicium*, into which were mixed chunks of leucitite. Atop this were erected two elliptical rings that served to delimit the upper element of the foundations. About 6 meters high, these also were made in *opus caementicium*, but with the addition of chunks of yellow tufa and with brick facing walls. The latter did not extend unbroken but were interrupted at the locations of the axial cryptoportici, or covered passageways, and various annexed spaces, thus forming four distinct quadrants that served as a permanent formwork for pouring the foundation's upper walls. This formwork was made with four superimposed layers of concrete that filled the entire volume of the quadrants delimited by the ring walls. Only at this point was the grade plane (Lat. *imum libramentum*) for the elevation of the construction reached, along with that for the floors.

ROMAN WALLS

With the spread of *opus caementicium*, Rome and the other cities of Latium and Campania developed specific construction methods associated with it. The *opus quadratum* of ancient tradition was adapted in the form of simple facing walls that were filled with *opus caementicium*, resulting in a permanent formwork made of squared blocks. Examples include the podiums (high masonry bases) of the great temples of Rome, from the late-republican example of the Temple of Portunus in the Forum Boarium to that of Antoninus and Faustina erected during the imperial period at the Forum Romanum, both made with a concrete core and a dressing in *opus quadratum*. On the Capitoline the great Tabularium, whose facade dominated the Forum Romanum, still preserves its high base in square blocks of Aniene tufa and peperino as well as part of its lower gallery, which has an arched facade made with blocks of peperino and travertine. Here also *opus quadratum* was associated with internal structures in *opus caementicium*. The same construction method was used for all the main public-entertainment structures in Rome, including the Theater of Marcellus, the Flavian Amphitheater, and the Stadium of Domitian. For the construction of such architectural types, other Roman cities also chose to combine the monumental appearance of *opus quadratum* with the versatility of concrete, as demonstrated by the amphitheaters at Capua, Pola, and Verona. The widespread use of this technique also involved structures for a variety of other purposes, such as triumphal arches, walls, and gates around cities, bridges, and aqueducts. The tomb of Caecilia Metella on the Via Appia, built early in the Augustan period, had a circular drum (29.15 meters in diameter, 11 meters high) raised on top of a high square base. The entire structure was made with strong concrete dressed in squared blocks of travertine. Despoiling of the lower base has extended over the centuries, its materials taken for use in other constructions, thus completely denuding the masonry structure and exposing technical aspects of its construction. The horizontal layers of the concrete filling are visible, as is the system of anchoring the dressing in *opus quadratum* by inserting bonding stones that remained partially incorporated in the *opus caementicium*. The upper masonry reveals a distinctive aspect of construction, as the surfaces of the facing

Above. Base of the tomb of Caecilia Metella on the Via Appia. Note the stratification of the opus caementicium *of the core, and surviving blocks of the facing wall made in* opus quadratum *of travertine blocks.*

Left. Opus quadratum *as a facing wall for* opus caementicium: *1. Orthostats alternating with stretchers; 2. Regular course of blocks arranged as headers and stretchers.*

Top. From the basilica of Paestum, detail of a stretch of wall in opus incertum *flanked by piers with semicolumns in* opus listatum.

Above. Circular tomb of the necropolis of the Porta Nuceria at Pompeii, dated to the first century B.C. Note the wall in opus incertum *and the thick layer of plaster decorated in false* opus quadratum.

blocks are rebated on three sides. In some cases these emphasized joints are not genuine but rather were drawn on the surface of the stone to simulate a perfect isodomic work.

Freed from the use of costly squared blocks, the Romans soon joined the methods of *opus caementicium* to other construction techniques that were more economical in terms of supply and application, including the construction of walls with concrete cores and facing walls made with small stone elements. In the second half of the first century B.C., Vitruvius (2.8.1) referred to the two principal construction methods then in use: "There are two kinds of walling; one like network, *opus reticulatum*, which all use now, and the old manner which is called *opus incertum*. Of these two methods the *reticulatum* is more graceful, but it is likely to cause cracks because it has the beds and joints in every direction. The 'uncertain' rough work, *opus incertum*, lying course above course and breaking joints, furnishes walling which is not pleasing but is stronger than the reticulatum."

As noted by Vitruvius, *opus incertum* was the older of the two building techniques. It called for the construction of facing walls made in irregularly shaped small or medium stones, either solid or broken. Construction usually took place by simultaneously raising both of the two outer facing walls (Lat. *crustae*) and the core between them, first using more regularly shaped stones along with a fairly thick mortar. For the concrete core the smallest stones were combined with waste materials, bound together to form a more liquid mortar that would penetrate every empty space. This technique was used to build walls in horizontal stretches, at variable heights. Sometimes Roman builders sought to accelerate the speed of construction using larger stones for facing walls, but this strategy presented the drawback of "burning up" the binding capacities of the mortar, as the stones would rapidly absorb the water in the mixture if they had not been sufficiently wetted. At Pompeii builders went so far as to draw the shapes of smaller elements on the faces of rocks, carving furrows in the shape of junctures and thus camouflaging the wall structure. To prevent such practices, various rules and guidelines were imposed on builders concerning the size of stones that could be used in walls. For example, the *lex puteolana parieti faciundo* (the "Puteolan Law on Wall Building") of 105–104 B.C. established that in the construction of a certain wall no stone could be used that weighed more than 15 *libra* (5.85 kilograms) when dry.

Opus incertum seems to have appeared in Rome as well as the other cities of Latium and

Campania in the third century B.C., with the first experiments in *opus caementicium*, and then spread elsewhere over the course of the next two centuries. Its use can be found in the large complex long associated with the Porticus Aemilia—although some scholars believe that it is related instead to the dockyard (*navalia*), also known as Porticus Aemilia, in the Roman port on the Tiber—and in Temple B on Largo Argentina, which was built around 100 B.C. The oldest and most monumental instances are found, however, outside Rome, beginning with several buildings at Cosa (modern Ansedonia) and Alba Fucens (Massa d'Albe), as well as the basilicas at Paestum and Pompeii and the sanctuaries at Praeneste (Palestrina) and Anxur Tarracina (Terracina). In cities the adoption of other building techniques beginning in the second half of the first century B.C. relegated *opus incertum* to structures of minor prestige, but in rural areas its use continued into later periods because the method was simple and economical to execute.

Toward the close of the second century B.C., Roman builders began experimenting with ways to regularize the construction of walls in *opus incertum*, and thus speed up the assembly of elements and allow for more rapid construction of facing walls. With the ensuing standardization of construction materials, stones arrived at the work site in the same shape and size, and could therefore be used without further dressing by stonemasons. Construction of a wall was reduced to the relatively less complex work of mounting and resulted in notable savings in time and cost. The savings of course depended on the large-scale availability of unspecialized slave labor that could continuously be put to work cutting stone at a quarry. Eventually, this process led to the development of *opus reticulatum*—undoubtedly the most "Roman" of all construction techniques and without equal in any preceding culture—in which outer facing walls were built from small diamond-shaped tesserae with a square base (measuring 6 to 10 centimeters on a side), which were laid facing out and with the pointed ends forming rows inclined at 45 degrees. There is nothing surprising about this method since it derives from the tradition of *opus incertum*, in which each stone was inserted in the space created above the juncture between two stones in the lower course. Thus the use of tesserae of the same shape made possible the creation of a true diagonal "network" with mortared joints, although by the time of Vitruvius, as we have seen, this type of construction was viewed as lacking resistance. Archaeologists, however, have reported the excellent condition of structures

Top. Opus reticulatum over the entryway to the Mausoleum of Augustus.

Above. Double wall of terracing for the monument built by Octavian (Augustus) at Nicopolis in Epirus to commemorate his victory at Actium. Note the facing wall in opus reticulatum and above it blocks with openings for the insertion of ships' rostrums.

Top. View of the amphitheater at Tarraco (modern Tarragona) in Spain. Most of the walls of the complex were made in opus listatum *using local stone.*

Above. Builders sometimes manipulated the fabric of walls in opus reticulatum, *creating polychrome compositions through alternating rows, even rendering figures or letters, as with this wall at Capua.*

built of *opus reticulatum* and the high quality of mortars used, often far stronger than the stone itself. The technique employed for the production of tesserae made use of material locally available, from various types of stone to broken pebbles and even broken bricks.

Among the most ancient examples of *opus reticulatum* are several buildings in Rome dating to between the end of the second and the beginning of the first century B.C., such as the basin of the Lacus Iuturnae at the Forum Romanum and Temple B at Largo Argentina. The technique grew in popularity over the course of the first century B.C., to triumph in the building programs of the emperor Augustus, including his house on the Palatine, his tomb (completed shortly after A.D. 28), and the theaters of Marcellus and Balbo, both erected at the Campus Martius. Perhaps it was from Rome itself that the technique spread to the entire area of central Italy, most especially Latium and Campania, dominating the construction of private and public buildings, as indicated by many structures at Pompeii and the other cities of Vesuvius, as well as by several monuments at Baia, Capri, Cassino, Cuma, and Liternum. In southern Italy *opus reticulatum* was adopted primarily at large construction work sites, such as the theater at Grumentum in Basilicata, the amphitheater of Lupiae (modern Lecce), and the theaters of Scolacium and Copia in Calabria. It also appears in Sicily, as for example in the amphitheater of Syracuse.

During the Augustan period *opus reticulatum* began to spread to the empire's provinces, but the technique saw only sporadic use in those areas. The most ancient evidence is the monument that Octavian built at Nicopolis in 29 B.C. to celebrate his victory over the ships of Antony and Cleopatra in the naval battle of Actium. Other examples can be found at Buthrotum (modern Butrint) in today's Albania near the Greek border, in Epirus and elsewhere in Greece (Athens, Corinth, and Olympia), in Asia Minor (Elaiussa Sebaste), Palestine (Caesarea and Jerusalem), and in northern Africa (Bulla Regia, Carthage, and Leptis Magna).

In many cases Roman builders used different kinds of tesserae for their *opus reticulatum*, creating polychrome compositions based on the various colors of the materials. On occasion different kinds of stone were used, and at other times stones were combined with tesserae made from cut-up bricks. The most widespread system was that of horizontally alternating courses, but in some cases builders arranged the tesserae to create letters and figures. The meanings of such compositions can be difficult to decipher today, especially since walls in *opus reticulatum* were made to be covered in plaster. However, many scholars do not rule out the possibility that, at least in some cases, the surfaces of walls were left uncovered. With the spread of fired bricks *opus reticulatum* began to decline (although it was used in the sumptuous villa built by Hadrian at Tibur, modern Tivoli, early in the second century A.D.), and disappeared altogether by the second half of the second century.

With the decline in use of *opus reticulatum*, another building technique began to spread in Roman architecture, this one based on the use of quadrangular blocks laid in horizontal courses and called *opus vittatum*, for its continuous horizontal courses similar to bandages (Lat. *vittae*). The archaeological record has demonstrated that it was used in different ways according to the geographic area: in the zones of diffusion of *opus reticulatum*, the use of *opus vittatum* was usually adopted to create angular chains of reinforcement for walls or to make specific structures, such as pylons for the conduction of water (at Pompeii, for example); in other contexts—Gaul most especially, but also Spain and northern Africa—it emerged as the dominant technique alongside *opus quadratum*, differing from that method only in the size of the blocks used. However, in Rome—after limited use around the middle of the second century in association with bricks—the technique only began to spread at the beginning of the fourth century A.D.

CONSTRUCTION TECHNIQUES USING BRICKS

The use of *opus caementicium* provided the impulse for the monumental development of Roman architecture during the imperial period, but it was fired bricks that served as the primary means for its advance. In reality, bricks were already part of the oldest building tradition in Rome, but generally in the form of mud bricks made with an argillaceous mixture and left to dry in the sun. This *opus latericium* was still being put to general use in construction in the first century B.C., with excellent results, at least according to Vitruvius (2.8.8–9), who emphasized the monetary value that Romans attributed to walls in mud bricks, as compared to those in stone: "If anyone will from these commentaries observe and select a style of walling, he will be able to take account of durability. For those of soft rubble with a thin and pleasing facing cannot fail to give way with lapse of time." To prove his point, Vitruvius refers to the valuation of rubble walls shared by

One of the brick columns from the basilica at Pompeii. Note the structure of the Attic base, made with shaped bricks covered in stucco.

CONSTRUCTION TECHNIQUES IN THE ROMAN WORLD

Opposite. Detail of the great exedra of Trajan's Market, the complex designed by Apollodorus of Damascus immediately northeast of Trajan's Forum. With the exception of travertine inserts, all of the monument was made using walls in opus testaceum *with a concrete core.*

Page 130, top. Detail of a wall in the temple at Liternum.

Page 130, bottom. Detail of a wall in opus mixtum *in the baths at Nicopolis. Visible are the accurate bonding courses in brick and the perfect toothing of the* opus testaceum.

two buildings: "Therefore when arbitrators are taken for party-walls [walls held in common], they do not value them at the price at which they were made, but when from the accounts they find the tenders for them, they deduct as price of the passing of each year the eightieth part, and so—in that from the remaining sum repayment is made for these walls—they pronounce the opinion that the walls cannot last more than eighty years." In contrast, brick walls do not require amortization: "There is no deduction made from the value of brick walls provided that they remain plumb; but they are always valued at as much as they were built for." Subsequently (2.8.17–18), however, he notes the deficiencies of brick walls for buildings of a great height: "Public statutes do not allow a thickness of more than a foot and a half to be used for party-walls. But other walls also are put up of the same thickness lest the space be too much narrowed. Now brick walls of a foot and a half—not being two or three bricks thick—cannot sustain more than one story." In fact, the population density of Rome renders the brick construction of homes untenable:

> Yet with this greatness of the city and unlimited crowding of citizens, it is necessary to provide very numerous dwellings. Therefore since a level site could not receive such a multitude to dwell in the city, circumstances themselves have compelled the resort to raising the height of buildings. And so by means of stone pillars, walls of burnt brick, party walls of rubble, towers have been raised, and these being joined together by frequent board floors produce upper stories with fine views over the city to the utmost advantage. Therefore walls are raised to a great height through various stories, and the Roman people have excellent dwellings without hindrance. Now, therefore, the reason is explained why, because of limited space in the city, they do not allow walls to be of sun-dried bricks.

At the beginning of the imperial period, perhaps in the wake of the standardization of construction materials that began with the tesserae of *opus reticulatum*, Roman builders paid particular attention to the possible uses of terracotta in wall construction. For centuries fired bricks had been used alongside trimmed tiles in the Magna Graecian cities of southern Italy. At Pompeii as early as the second half of the second century B.C., the columns of the basilica were made using fired bricks specially shaped to create wide grooves. The Romans immediately understood the advantages offered by bricks, including the low cost of their production, which was connected to the use of widely available raw materials (clay, sand, and straw), the speed of their installation, and their excellent strength.

Whole bricks were rarely used in masonry, and most often reduced to smaller triangular or rectangular elements (Lat. *semilateres*) by breaking or sawing the brick along the transversal axis or the diagonal. With these segments masons made facing walls in *opus testaceum*, while the hollow of the core was filled with a solid *opus caementicium*.

The use of fired bricks began to take hold in Rome early in the first century A.D. Its initial uses in monumental architecture arose during the reign of Tiberius, when they were employed for the construction of his palace on the Palatine Hill as well as that of the barracks for the Praetorian Guards (Castra Praetoria) northeast of the city. From that time forward the use of *opus testaceum* spread among the city's work sites, and by the time of Nero's reign was established as the principal building method. The Domus Aurea, Nero's gigantic residence, was built in fired bricks as was, later, the great palace of the Flavian emperors on the Palatine. Also using *opus testaceum* were the monumental constructions built under Trajan and designed by the architect Apollodorus of Damascus, including the Forum of Trajan, the markets erected along the slopes of the Quirinal, and the baths on the Oppian Hill.

Also constructed in brick were the giant bathing complexes built by Diocletian and Caracalla and, finally, the great projects of the late empire, such as the Aurelian Walls and the Basilica of Maxentius.

The technique rapidly spread to all the empire's cities. There is probably not a single site of importance from the imperial period where fired bricks were not used. Homes, temples, structures for spectacles, city gates, fountains, aqueducts—all types of architecture could be made in bricks. Examples include the numerous homes and the Capitolium at Ostia, the theater and odeion at Nicopolis in Greece, the Porta Palatina at Turin (originally Augusta Taurinorum), the baths at Cluny in France, and the interminable succession of arcades in the aqueduct built by Alexander Severus in Rome (Aqua Alexandrina).

THE COMBINATION OF WALL TYPES

As indicated by Vitruvius's passage on *opus incertum* and *opus reticulatum*, the Romans were well aware of the qualities of various types of masonry techniques and their relative strengths. That understanding, along with solutions adopted at work sites to simplify the construction of walls, soon led to the combination of different types of materials as well as different methods of their use. The result was *opus mixtum*, a term that has come to encompass several distinct building methods.

The oldest evidence for this technique dates to the late second and first centuries B.C., when the use of facing walls, first in *opus incertum* and then in *opus reticulatum*, led to the need for standardizing and reinforcing walls with sharp corners. In many cases, as in the monumental vaulted substructures at the Sanctuary of Jupiter Anxur at Tarracina (early first century B.C.), recourse was made to medium and large stone blocks; in others, such as the odeum at Pompeii (around 80 B.C.) or the theater at Cassino (Augustan period), preference was given to the use of corner supports built with bricks. These instances, however, presented the necessity to connect the wall structure to a fabric composed of *opus reticulatum* that was installed in rows inclined at 45 degrees. As a result two different systems of toothing developed; the first used saw teeth that followed the outlines of the tesserae of the reticulatum, while the second used square teeth and was better suited to the horizontal arrangement of courses associated with the techniques of *opus vittatum* and *opus testaceum*.

In other instances, *opus mixtum* was distinguished by a combination of different masonry techniques not only in the corner areas but over the entire extent of the walls. The most widespread example was presented by panels in *opus incertum*, *reticulatum*, or *vittatum* that were framed by horizontal bonding courses and vertical bands of *opus testaceum*. This system provided reinforcement by improving the anchorage of facing walls to the concrete core, and also offered masons the advantage of working on regular horizontal planes during construction.

Opus mixtum involved not only the combination of different techniques but also the combination of different materials, as represented by *opus vittatum mixtum*, which alternated courses of stone blocks and fired bricks for the installation of facing walls. Present at Pompeii as early as the first century B.C., this technique spread throughout the Roman world primarily during the middle- and late-imperial periods, and can be found at the large complex built by Maxentius early in the fourth century A.D. along the initial stretch of the Via Appia. Structures there included the imperial palace as well as a circus and the tomb known as the Temple of Divus Romulus, from the name of Maxentius's son, who was buried there.

ARCHES IN *OPUS QUADRATUM*

In the first century A.D. the Roman philosopher and statesman Seneca (*Epistles* 90.32) related, and dismissed, one account of the origin of the arch: "Posidonius again remarks, 'Democritus is said to have discovered the arch, whose effect was that the curving line of stones, which gradually lean toward each other, is bound together by the keystone.' I am inclined to pronounce this statement false. For there must have been, before Democritus, bridges and gateways in which the curvature did not begin until about the top." Here Seneca refers to the Greek philosopher Democritus of Abdera, who lived between the fifth and fourth centuries B.C. Because of his belief that all things are composed of atoms, Democritus is sometimes considered the father of modern

View of the arcaded facade of the Theater of Marcellus in Rome.

The parts of an arch in blocks.

Opposite. System of arcades in travertine blocks at Porta Maggiore in Rome.

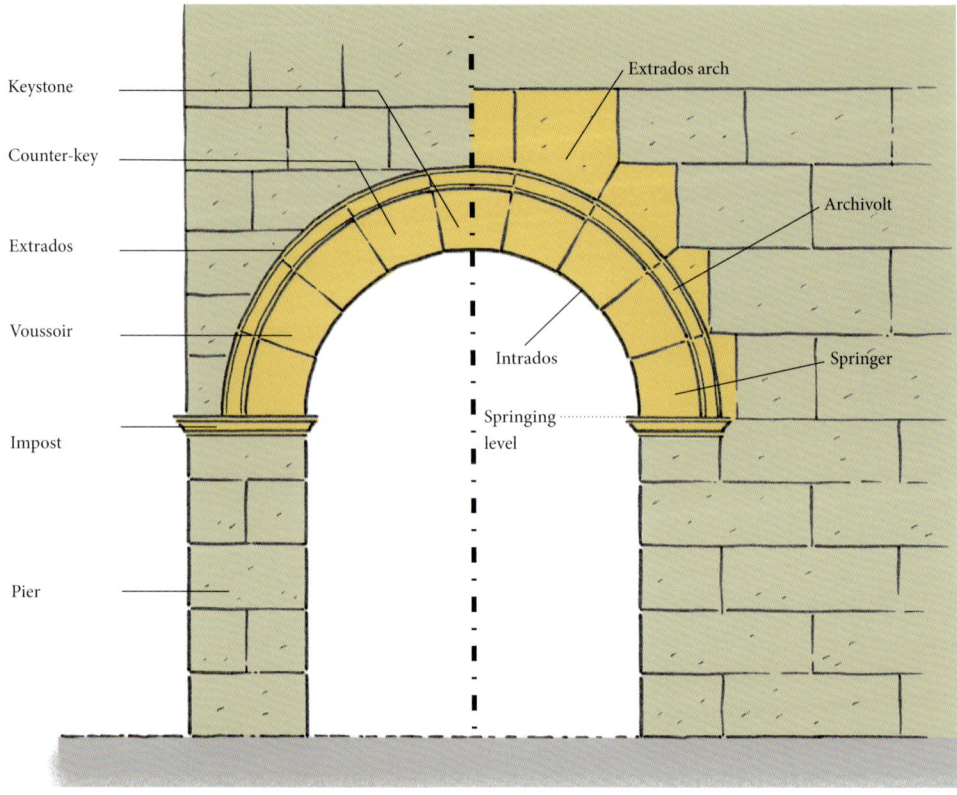

physics. Posidonius, the Greek philosopher born at Apamea in Syria around 130 B.C., credited Democritus with the invention of the arch; but Seneca clearly had doubts concerning this hypothesis. Archaeological research has in fact established that both the Mesopotamian and Egyptian cultures built arched structures as early as the fourth or third millennium B.C. A fine example is the Temple of Tell al-Rimah in modern Iraq, which was built around 2000 B.C. with a series of vaulted spaces up to 3.8 meters wide, as well as a stairway supported by arches. Arched structures spread in the Greek and Magna Graecian world only from the fourth century B.C., for monumental architecture as well as minor buildings, such as chamber tombs.

The arch (Lat. *fornix* or *arcus*) is a curved structure usually composed of several elements arranged radially around a center and set upon two uprights, which receive the weight of the radial elements along directions that diverge from the vertical. An arch can be built to occupy the entire thickness of the wall in which it is inserted, and in this sense it differs from a vault (Lat. *confornicatio* or *camara*), which instead covers an entire area. The most elementary form of an arch (Lat. *arcus cuneis conclusa*) is composed of wedge-shaped stones called voussoirs (Lat. *cunei*), a series of wedge-shaped elements that converge on a central piece, the keystone of the arch. The keystone is of fundamental importance since it prevents the tendency of voussoirs to collapse toward the inner space, as Seneca (*Epistles* 118.16) notes in another passage: "Some things endure according to their kind and their peculiar qualities, even when they are enlarged. There are others, however, which after many increments, are altered by the last addition; there is stamped upon them a new character, different from that of yore. One stone makes an archway—the

CONSTRUCTION TECHNIQUES IN THE ROMAN WORLD 133

Arch of Constantine, Rome.

Pons Aemilius that span the Tiber date to a few decades later. Other information about ancient arched structures comes from literary sources. Livy (*History of Rome* 33.27.4–5) recalls that, already in 196 B.C., C. Lucius Stertinius had erected one arch in the Circus Maximus and another two in the Forum Boarium, in front of the temples of Fortuna and Mater Matuta. In all these cases arches were constructed with stone blocks, a technique that remained in use through later centuries, as demonstrated by the arches of the amphitheater at Capua, dated from the end of the first to the second century A.D.

As we have seen, in the late second century B.C. and particularly in the following century, the spread of the use of concrete at Roman work sites led to a profound revolution in the construction arts, inaugurating a trend that quickly produced a completely new way of designing the internal spaces of buildings. Just as walls made of squared blocks came to serve as mere dressing for a concrete core, the use of squared stones in arches and vaults was reduced to defining only the outer facades of such structures. This technique had already been adopted in the great arcades and substructures of complexes built in Latium during the Late Republican period, including the Sanctuary of Jupiter Anxur at Tarracina and that of Hercules Victor at Tibur. In Rome in 78 B.C. the architect L. Cornelius Catulus built the Tabularium, an impressive structure that held the state archives and at the same time supported the eastern slopes of the Capitoline. Its facade, facing the Forum Romanum, was composed of two orders of arcades in blocks of peperino framed by half-columns. Behind the Tabularium opened a long covered hall with cloister vaults in *opus caementicium* and a series of other spaces, all covered in concrete vaults. This association of *opus quadratum* with arches, and of *opus caementicium* with internal structures, continued until the middle of the imperial period, and eventually characterized all the great buildings for spectacles

stone which wedges the leaning sides and holds the arch together by its position in the middle. And why does the last addition, although very slight, make a great deal of difference? Because it does not increase; it fills up."

The oldest arched structures in Rome appear to date to the late third and second centuries B.C. In the Forum Romanum eight arches of small tufa blocks were built in 179 B.C. to support the Clivus Capitolinus, the road that climbed the Capitoline to the Temple of Jupiter Capitolinus. The large arches of the

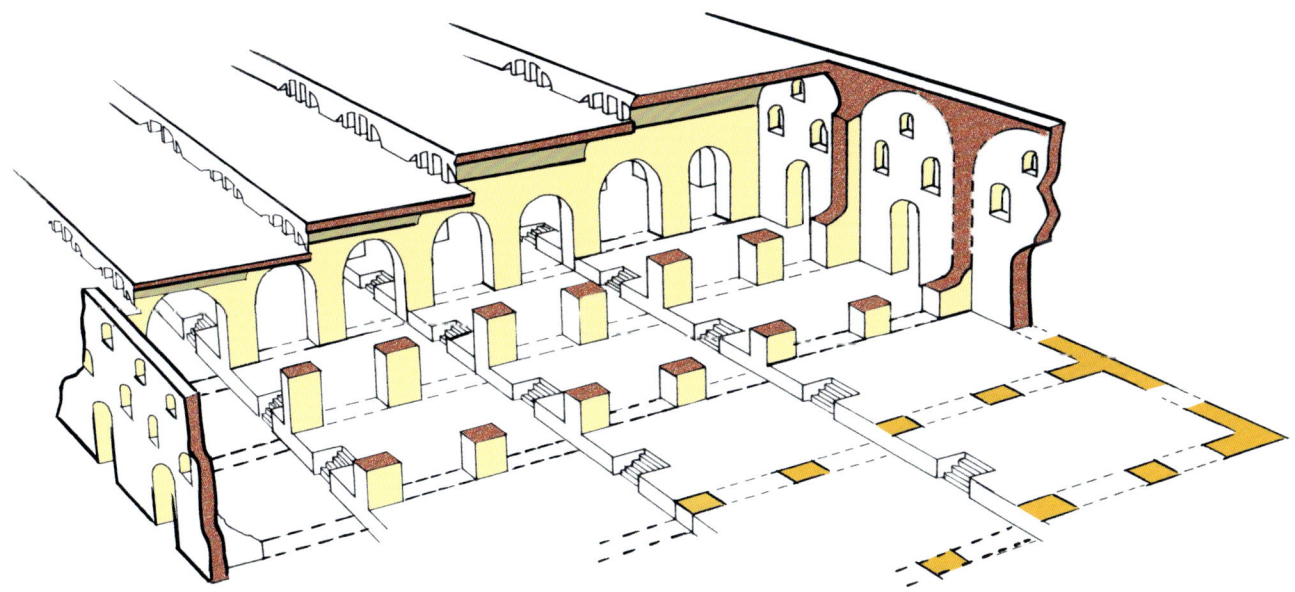

Reconstruction of the building traditionally identified as the Porticus Aemilia, built in Rome in 193 B.C.

both in Rome (the Theater of Marcellus, the Colosseum, and the Stadium of Domitian) and in provincial cities (modern Arles, Catania, Lecce, Pola, Tarragona, and Verona).

In Rome and more generally in the western part of the empire, with the introduction of concrete only the outer faces of arches and vaults continued to be made of stone blocks. However, in the eastern area the ancient tradition of building in stone was carried on to produce large structures with stone-block arches. Examples include numerous buildings erected in Asia Minor, such as the nymphaeum at Side and the gymnasium and bath complex at Termessos.

Until now we have spoken of the most common shape of the arch, that of the round arch, whose outline follows a half-circumference traced from a center located along the impost line. But the Romans also made use of the particular type of arch today called a flat or jack arch, or even a platband. This was indeed a flat arch, similar in shape to an architrave, but composed of several voussoirs cut so that they converged toward the center. Such an arrangement offers the advantage of producing an architrave that is not monolithic. Its function was similar to that of a round arch, in the sense that every voussoir transmitted its lateral thrust to the adjacent voussoir, and the entire system was held in place by a central keystone. In Rome flat arches in *opus quadratum* were used as early as the first century B.C. for the Doric-order architrave of the Tabularium, and later in the two superimposed levels of the shops at the Forum of Caesar. For the Temple of Castor and Pollux at the Forum Romanum, and for the Porta Maggiore, the flat arch was used not as an architrave but rather in association with one, thus performing as a true relieving arch. There are cases in which a flat arch is reinforced by the insertion of specially shaped metal rods. This solution, attested in the porticoes of the forum of Pompeii and in some of the buildings at Hadrian's Villa at Tibur, notably strengthened the structure by reducing the stress of traction that was present in the lower part.

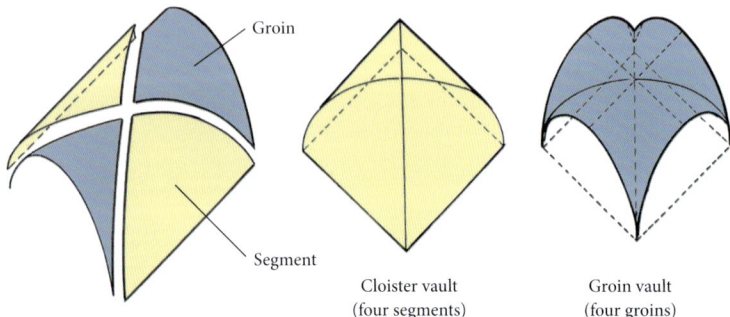

Top. Basilica of Maxentius, Rome.

Above. Methods of forming vaults composed of segments (cloister vault) and groins (groin vault).

ARCHES AND VAULTS WITH CONCRETE CORES

The application of *opus caementicium* to arched and vaulted structures required a variety of technical solutions and resulted in a wide range of forms. By the second century B.C. the Romans had begun exploiting opportunities it offered for the creation of ceilings and roofs. Due to its characteristic semifluidity in the installation phase, *opus caementicium* had the capacity to assume whatever shape was established by the formwork (or, more accurately, by the centering erected for construction of the vault); when the mortar had set it was transformed into a structure with elasticity and strength, comparable to a monolith. However, all stages of this process depended on the quality of the mortar used and on the masons' skillful calculation of internal tensions and counterthrusts, as well as their expertise in positioning the aggregates.

With vaults, as with arches, the simplest shape used by Roman architects was the round. This form was also the most widespread, used in every building type. All vaults were made with wooden centering, into which concrete was poured. The building traditionally considered the most ancient application of concrete vaults on a monumental scale is the Porticus Aemilia, a large complex (487 by 60 meters) erected in 193 B.C., and divided into fifty parallel corridors (each 8.3 meters wide) covered by a series of vaults. Beginning with their first experiments, Roman builders—perhaps mindful of the knowledge gained through the creation of arches in blocks—adopted the custom of radially arranging stones or bricks destined to be incorporated in the *opus caementicium*, at least on the intrados, or inner curve, of the structure. This solution was already in use between the end of the second and early first centuries B.C. for the vaults of the Sanctuary of Fortuna Primigenia at Praeneste, and ensured that, in the case of sagging, cracks would develop in radial directions, forming areas of the vault that resembled wedges.

For both vaults and arches, concrete was often used only for the internal filling of a structure, in a manner similar to that employed for facing walls. This technique appeared most frequently in triumphal arches and aqueducts, where the *opus quadratum* of the outer faces and archivolts also characterized the entire intrados. Examples include the Arch of Titus and that of Constantine in Rome, as well as that of Trajan at Ancona, and the Pont du Gard, built in the Augustan period to bring water to Nemausus (modern Nîmes) in France. In other cases concrete was poured directly into the centering, and archivolts were applied only to the facade, which was often made of bricks arranged in a single row, or in many rows, with double or triple archivolts.

Restraining arches (Lat. *fornicatio*) represent a special case. This type of arch is inserted in a wall to protect the part beneath it by directing thrusts outward and around the lower section. In general, restraining arches are located above the architrave of a door or the lintel of a window. A complex system of superimposed restraining arches was adopted for the Pantheon, the archivolts of which are visible on the external faces of the walls.

The field of compound vaults includes all those with intrados surfaces formed from the intersections of different segments. The basic components of a compound vault are the groin and the rib, both generated from the division of the round arch along its diagonals. The assembly of four groins creates a cross, or groin, vault, as in the vast central aisle of the Basilica of Maxentius. The intersection of four ribs forms a cloister, or tent, vault; an early application dating to the first century B.C. is found in the gallery of the Tabularium in Rome. For more-complex compositions Roman architects multiplied the number of groins and ribs, creating coverings with hexagonal, octagonal, and even decagonal bases, such as the large dome of the Temple of Minerva Medica in Rome. The Roman consul Dio Cassius (*Roman History* 69.4.2) recalled that Apollodorus of Damascus, architect to the emperor Trajan, was once annoyed with the young Hadrian, who enjoyed making architectural designs, and told him: "Be off, and draw your gourds. You don't understand anything of these matters." When Hadrian became emperor in A.D. 117, he exiled Apollodorus and created the great residence known as Hadrian's Villa at Tibur, composed of scenographic spaces and areas covered by daring vaults.

TRAJAN'S BATHS: TRACES OF AN ANCIENT "TIME SHEET" AT A ROMAN WORK SITE

Constructed partly on top of the remains of Nero's Domus Aurea, the baths built by the emperor Trajan and erected in only five years—from A.D. 104 to 109—constituted the first example of the "great baths" in Rome. The complex occupied an area of 315 by 220 meters and was organized by a large rectangular enclosure, with porticoes outside and exedrae inside that extended the area of the actual baths (212 by 190 meters). Based in part on the account of

Arcade of the amphitheater at Capua (late first–early second centuries A.D.), framed by an order with an architrave composed of a flat arch made of stone blocks.

Among the few remains of the monument, spread on the Oppian Hill, is a long subterranean gallery inside the western substructures. The gallery, 8 meters wide and covered by a depressed round arch (15 meters high), was built of brick walls with concrete cores on a reinforced-trench foundation of *opus caementicium*. During construction, workers encountered older structures, among them a large building from the Flavian period with arched openings and decorations in painted plaster, including a beautiful depiction of a bird's-eye view of a city. The arches were filled in with brick walling during the Trajanic period at the time of the gallery's construction.

Without doubt the most interesting aspect of this gallery is the presence of a series of twenty-eight inscriptions painted in red on the face of the wall. These letters indicate dates that are, of course, expressed in the system of the Roman calendar. For example, there is "XI K MA," which means "[*die*] XI [ad] K[*alendas*] MA[*ias*]" (eleven days before the kalends of May), and corresponds to the modern era's April 21. These inscriptions, the only ones of their kind within the archaeological panorama, constitute an interesting testament to the organization of a work site during the Roman period and the methods used to record the daily activities of masons. The inscriptions in fact refer to the portion of wall built during each day of work, and were marked there at the end of the day. On the basis of these inscriptions it has been possible to reconstruct, step by step, all the phases in the construction of the wall as well as the double-archivolt arch inserted in it. We know that construction of the arch began on April 2 and was not completed until April 17, resulting in two full weeks of work. Furthermore, the date in the inscription indicated as April 21 corresponds to the celebration of the *Paliliae*, a religious festival commemorating the day of Rome's founding, thus indicating that at least some of the masons were required to work on that day, a holiday for others.

Dio Cassius (*Roman History* 69.4.1), the design of the monument has been attributed to Apollodorus of Damascus, the architect of Syrian origin who also created Trajan's Forum and the nearby markets. Very little remains of the great bathing complex, but it has been possible to reconstruct its plan by reference to the *Forma Urbis Romae*, the marble plan of Rome created under the emperor Severus.

ENGINEERING AND TECHNIQUES AT THE WORK SITE

FROM THE QUARRY TO THE WORK SITE: SOME TECHNICAL METHODS OF THE GREEK WORLD

Following extraction from the quarry face and rough shaping, every block of stone or marble began a trip (Gr. *lithagogia*) to the work site for which it had been prepared. Maneuvering these elements, each weighing several tons and some in shapes that made them relatively fragile, was not an easy operation. To confront these issues, the Greeks devised several truly ingenious methods of transport.

Some quarries, such as those on the island of Thasos, were located along the coast, making it easier to load pieces onto ships for transportation by sea (Gr. *lithagogia kata thalassa*). An inscription from the Hellenistic period relating to the reconstruction of the temple of Apollo at Didyma recalls one remarkable system used for the maritime transportation of blocks: the larger ones were lashed with cables to wooden boards held crosswise between two double-bow ships (Gr. *amphiprymnoi*). The stone pieces thus traveled immersed in water, the force of which lightened the load.

Quite often, however, maritime transportation involved only one of the stages in the journey to be taken by stone material. For example, blocks of tufaceous limestone quarried in the area of Corinth and destined for the construction of a building at Delphi could be loaded on ships (Gr. *lithagogi*) and transported by sea only as far as the port of Kyrrha, from where

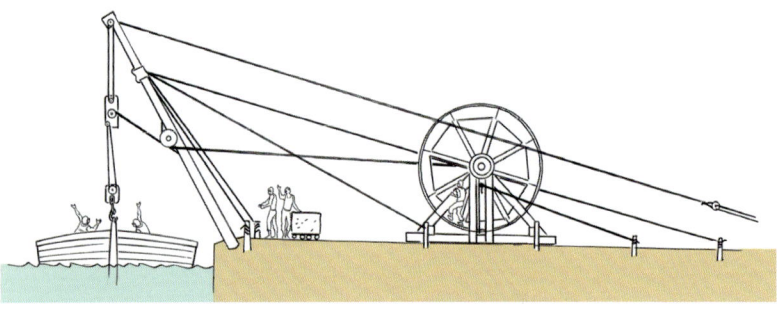

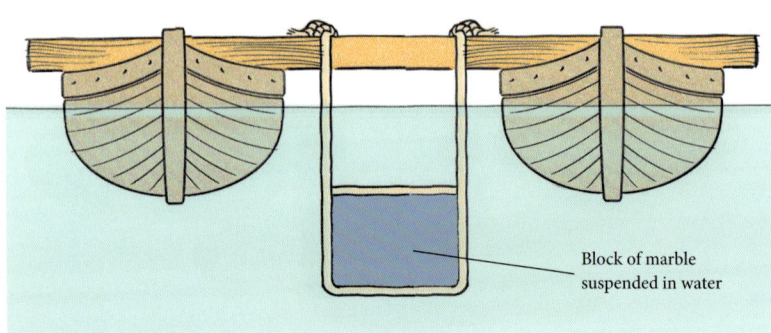

Block of marble suspended in water

they would continue by land to the Sanctuary of Apollo, located at an altitude of 650 meters above sea level.

Other quarries were situated further inland, often at relatively elevated sites, making transportation of stone blocks even more arduous. Beginning in the fifth century B.C. white marble was quarried on Mount Pentelicon to build the principal monuments of Athens. Several quarry faces have been identified, since they were opened as needed when major construction was launched in the city. Among these work sites were the Parthenon and the Propylaea on the Acropolis, as well as the Temple of Athena and Hephaistos (the Hephaisteion) near the agora, in the Classical period; the Stoa of Attalus II in the agora and the resumption of work on the great Temple of Zeus Olympios in the area of the Ilissus, in the second century B.C.; and completion of the architectural complex of the Temple of Zeus and construction of the stadium, in the second century A.D.

Manolis Korres, an archaeologist who has devoted many years to studying the largest temple dedicated to Athena on the Acropolis, has reconstructed the route taken by Pentelic marble from the quarry to urban work sites. On the first stage of the trip, blocks traveled on roads that had been built at an inclination of 30 percent and specially paved to act as slipways. The blocks were put on a sort of wooden sledge (Gr. *eschara* or *chelone lithegos*), whose movement was guided by means of levers. To slow their descent the sledges were held by cables anchored to bitts, each with a diameter of about 30 centimeters, that were fixed to the sides of the road. When the sledges reached the foot of the mountain, the blocks were loaded on a cart (Gr. *tetrakyklē*) drawn by animals, and by that means traveled the roughly 17 kilometers from the slopes of Mount Pentelicon to the city. From that point the cart was required to make the steep ascent to the citadel, via the western access road in the direction of the Propylaea. The difficulties

Page 139. View of the quarry for white marble at Aliki on Thasos. Traces of ancient step quarrying are visible, as is the site's easy access to the sea.

Opposite, top. Reconstruction of the operation by which a block of white marble from the quarry of Aliki on Thasos was loaded onto a ship.

Opposite, center. An inscription related to the construction of the Temple of Apollo at Didyma relates that large pieces were transported by sea, suspended in water.

Opposite, bottom. Reconstruction of the system of balanced carts used to transport blocks of Pentelic marble on the Acropolis.

Left. High quarry face of Pentelic marble near Spilia. Still visible are numerous ancient cut marks.

posed by the climb were resolved by using a system of balanced carts: the cart carrying the marble was drawn by cables that passed through a pulley (Gr. *trochilos* or *manganon*) fixed at the highest point; these cables were tied to a second cart, drawn by animals, that descended in the opposite direction. The quarried material could thus reach the plateau of the Acropolis and the construction site.

Transportation costs for stone varied according to the means of travel and, of course, according to the distance. An inscription related to limestone blocks from Corinth destined for the sanctuary at Delphi indicates that cornerstones could be purchased at the quarry for 61 drachmas, but the price increased dramatically with the expense of transportation: 240 drachmas for the sea route, increasing to 420 drachmas for travel by land from the port of Kyrrha to the sacred area. Another inscription relates to the transportation of Pentelic marble to make colonnades for the great hall (*telesterion*) of the Sanctuary of Demeter at Eleusis, where the Eleusian Mysteries were performed. Renting oxen to pull the carts cost an average of 4 drachmas and one-half obolus, and the transportation of a single-column drum required from two to two-and-a-half days of travel. Added to these costs were expenses for the workers employed in the transportation, from the man driving the cart to the assistants (Gr. *zeugetrophoi*) who cared for the animals. Travel by sea was less expensive, as indicated by an inscription relating to the Sanctuary of Asclepius at Epidaurus in Argolis, which gives the cost for transporting a block of Pentelic marble from the Athenian port of Piraeus to the port near Epidaurus: just 25 drachmas.

Whether stone was drawn on carts or traveled in special ships, the most complex operation was simply moving it onto the chosen mode of transportation, since each block could weigh several tons. Experience soon led

to the creation of large hoisting machines. At Thasos, for example, blocks of white marble were lifted and loaded on ships by means of a crane driven by a large wheel that rotated through the movement of men who walked inside it.

In some cases architects devised specific technical solutions that transformed the architectural pieces themselves into mobile systems. Vitruvius (10.2.11) describes "the ingenious contrivance of Chersiphron" used during construction of the great Artemision of Ephesus:

> When he desired to bring down the shafts of the columns from the quarries to the temple of Diana at Ephesus, he tried the following arrangement. For he distrusted his two-wheeled carts, fearing lest the wheels should sink down in the yielding country lanes because of the huge loads. He framed together four wooden pieces of four-inch timbers: two of them being cross pieces as long as the stone column. At each end of the column, he ran in iron pivots with lead, dovetailing them, and fixed sockets in the wood frame to receive the pivots, binding the ends with wood cheeks: thus the pivots fitted into the sockets and turned freely. Thus when oxen were yoked and drew the frame, the columns turned in the sockets with their pivots and revolved without hindrance.

The cylindrical shape of the shafts permitted Chersiphron to develop a system that converted the material to be transported into a long wheel, but this strategy would not serve for the movement of other architectural elements such as an architrave, with its characteristic elongated parallelepiped shape. However, workers were able to proceed, thanks to the efforts of Metagenes, the son of Chersiphron, who adapted the method used to convey the shafts to the transport of the lintels, as Vitruvius (10.2.12) describes: "He made wheels about twelve feet in diameter, and fixed the ends of

View of the Parthenon, built in the second half of the fifth century B.C. entirely in marble from Pentelic quarries.

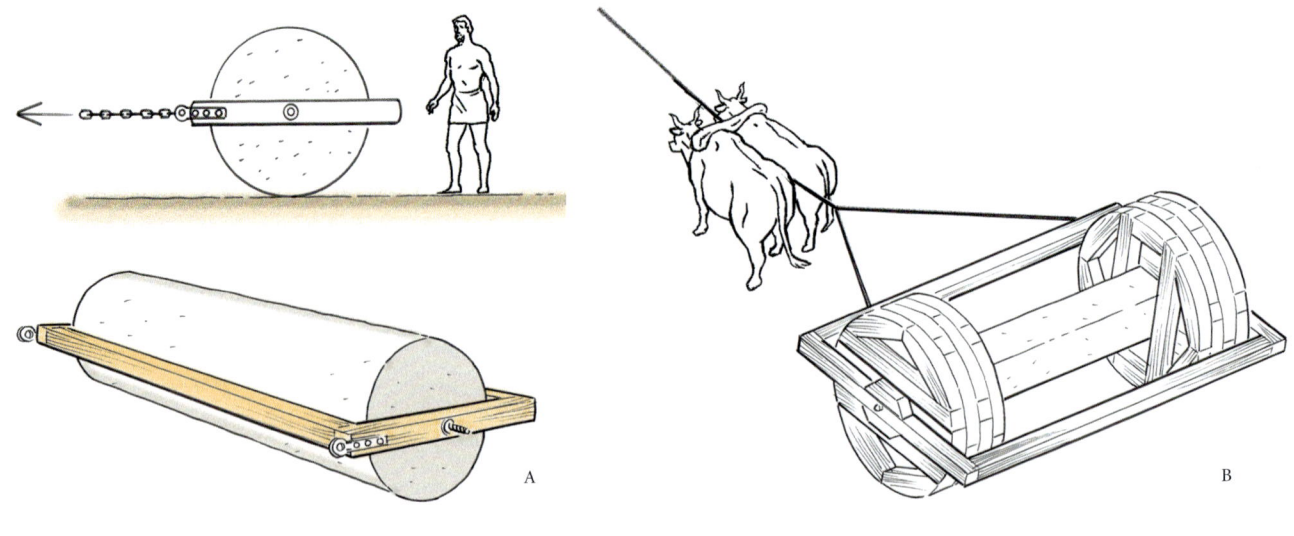

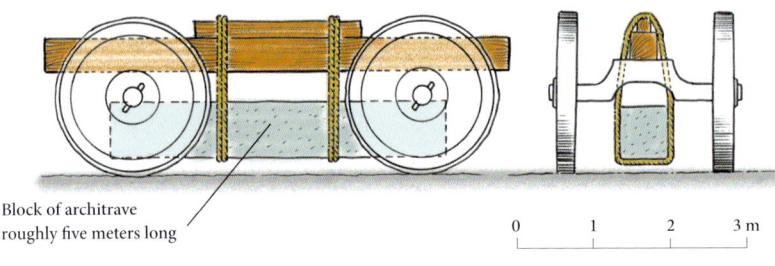

Block of architrave roughly five meters long

0 1 2 3 m

Top. Systems devised in the sixth century B.C. by Chersiphron (A) and his son Metagenes (B) to transport large architectural elements for the Temple of Artemis at Ephesus.

Bottom. Long stone blocks could also be transported by tying them with cords to the axles of a large cart.

the architraves in the middle of the wheels. In the same way he fixed pivots and sockets at the ends of the architraves. Thus when the frames of four-inch timber were drawn by the oxen, the pivots moving in the sockets turned the wheels, while the architraves being enclosed like axles in the wheels (in the same way as the shafts) reached the building without delay." As Vitruvius notes, "This expedient would not have been possible unless, to begin with, the distance had been short. It is not more than eight miles from the quarries to the temple, and there are no hills but an unbroken plain."

The methods devised by the two architects active in the Ephesian sanctuary find an interesting archaeological mirror on the work site of the gigantic Temple G at Selinus. This building's architraves, truly colossal in size (more than 6.5 meters long and 2.3 high), preserve the special grooves made for their transfer from the quarry at Cusa. This travel was probably made possible by constructing a horizontal wooden framework around each piece, to which the hubs of large wheels were attached; for all intents and purposes the block became the body of the cart itself.

ENGINEERING AND MECHANICS: HOISTING MACHINES

The most complex operation in the construction of a stone building was that of raising individual architectural elements, often to extraordinary heights, and positioning them. The large blocks of the trabeation for the Temple of Apollo at Didyma in Asia Minor stand at a height of about 20 meters from the level of the stylobate, while the travertine blocks on the facade of the Colosseum extend to a height of nearly 50 meters. To construct monuments of such dimensions the Greeks invented a series of hoisting machines that were later adopted and further elaborated by the Romans. In brief, these machines were based on the application of three different mechanisms: the pulley (Gr. *trochilos*, Lat. *orbiculus*), the windlass (Gr. *stropheion*, Lat. *sucula*), and the block and tackle (*trochileia*). The pulley, probably invented by Greek traders, was composed of a simple wheel with a guide groove for the cable. One end of the cable was attached to the block, and the cable was then pulled from the opposite end to lift the block; however, this system required the application of a force equal to the weight of the element being lifted. The windlass served to reduce the effort of lifting by wrapping the cable around a movable drum, so that the turning of the drum drew the end of the cable. Even greater advantage could be gained through the use of a block and tackle, composed of several connected pulleys (Gr. *auchenes*, Lat. *trochleae*). This system made it possible to reduce the force needed in proportion to the number of pulleys: two pulleys divided in half the force that was required, four lowered it by a quarter.

Vitruvius describes the principal types of hoisting machines in detail and explains their operation. First, however, he introduces (10.1.1) the definition of a machine, "a continuous material system having special fitness for the moving of weights. It is moved by appropriate revolutions of circles, which by the Greeks is called *cyclice cinesis*. The first kind of machine is of ladders (in Greek *acrobaticon*); the second is moved by the wind (in Greek *pneumaticon*); the third is by traction (in Greek *baru ison* or *equilibrium*)." Vitruvius further distinguishes two types of machines: "Machines of draught draw weights mechanically so that they are raised and placed at an elevation. . . . Traction machines offer in practice greater adaptation which reaches magnificence, and when they are handled carefully, supreme excellence." He then describes (10.2.1–3) the construction and mechanics of the simplest type of lifting machine:

> Two pieces of timber are carefully prepared, which answer to the size of the load. They are set up, connected at the top with a brace, and spreading at the base. They are kept upright by ropes fastened at the top and adjusted around them. At the top a block is made fast: these some call *rechamus*. On this block two pulleys are fixed, which revolve upon axles. Over the top pulley the leading rope is passed. It is then let down and drawn around a pulley of the block below. It is returned to the lower pulley of the top block, and so comes again to the lower block and is secured to the eye of it. The other end of the rope belongs to the lower part of the machine.
>
> On the back faces of the timbers where they separate, socket pieces are fixed, into which the ends of the windlasses are put, so that the axles may turn easily. The windlasses near their ends have two perforations so adjusted that handspikes can fit into them. To the bottom of the block, iron pincers are fixed, the teeth of which are adjusted to holes in the blocks of stone. Now when the rope has its end tied to the windlass, and the handspikes draw and turn the windlass, the rope in winding around the axle is made taut and so lifts up weights to their place in the work.
>
> Now this kind of contrivance, because it is turned by three pulleys, is called *trispastos*. When, however, there are two

A crane powered by men inside a wheel, as depicted on a bas-relief by Lucceius Peculiaris found at Capua.

ENGINEERING AND TECHNIQUES AT THE WORK SITE 145

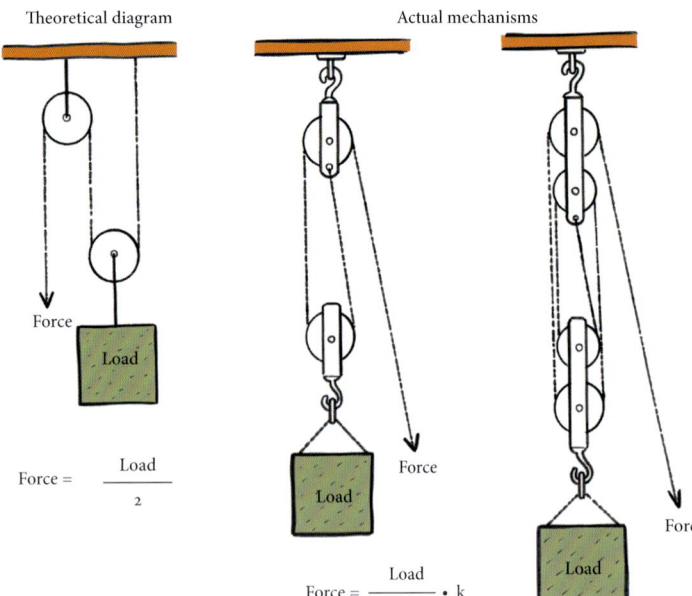

Principles behind the operation of a block and tackle.

pulleys in the lower block and three in the upper block, it is called *pentaspaston*.

What Vitruvius describes is in fact a crane, a system of lifting that is depicted in several images from the Roman world and still used today. A crane appears in a clay relief found along the Via Cassia that illustrates the foundation of a city, as well as in a marble slab from Anxur Tarracina (modern Terracina) that reproduces a scene of construction at that city's port. Even more detailed is the image of a hoisting machine included in the construction scene painted in the *caldarium* of the Villa of San Marco at Stabiae. In this view two people are shown operating the machine while a third, standing on top of the wall being constructed, awaits the arrival of the block to lay in place.

"If machines are to be prepared for greater loads," advises Vitruvius (10.2.3–4), "we must use longer and thicker timbers. In the same way we must use larger bolts at the top, and larger windlasses below." When everything is ready the tackle, which had previously been loose, was attached, and the cables were carried over the shoulders of the machine. "If there is no place to which they may be fixed," he stipulates, "sloping piles are to be driven into the ground and secured by ramming the ground around them; to them the ropes are to be attached." At the top of the machine the block is attached by a cable, and a rope is extended and secured to the block located on top of the inclined pile. The rope is passed over its pulley and carried back to the block bound to the top of the machine. The rope then descends and returns to the windlass below, where it is bound. As a result, concludes Vitruvius, "the windlass being worked by handspikes will revolve: and of itself will raise the machine without danger. Thus the ropes are passed around, the cables are fixed to the piles, and the machine is in position for use. The pulleys and the tackle are applied as it is described above."

But when the dimensions and weight of the material to be moved are great, Vitruvius (10.2.5–7) cautions, one "must not trust to the windlass." Instead he recommends the insertion of "an axle held in sockets like the windlass . . . having in the middle a large drum, which some call a wheel: the Greeks, *amphieres*, or otherwise *perithecium*." The blocks for these machines are made differently than those for a crane: "For they have below and above two pulleys arranged vertically. The guide rope passes into a hole in the lower block in such a way that the rope when it is taut has its two ends equally long." The rope is then passed around and fastened to the lower block, and both ends of the rope are secured so that they cannot swerve to the right or the left. The ends of the rope are taken back along the outside of the upper block and over its lower pulleys, returning below. At this point, the ropes are passed from the inside to the pulleys of the lower block and then carried up on the right and the left, returning to the top around the highest pulleys. Passing from the outside they are carried on either side of the drum on the axle, and are tied securely. A second rope is then wound around the drum and carried

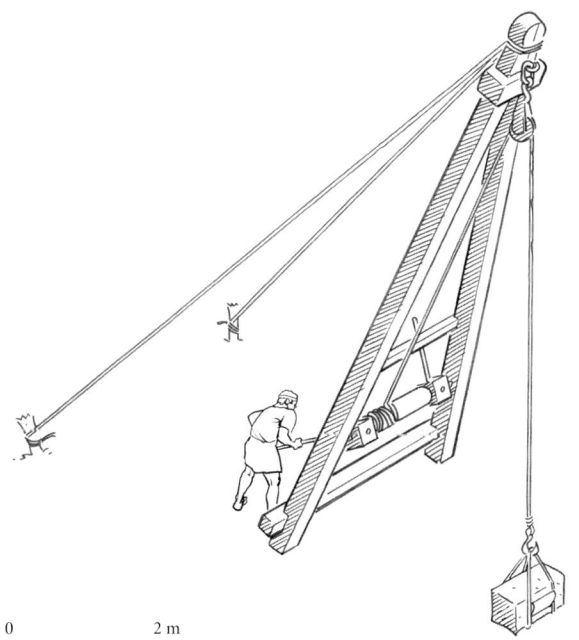
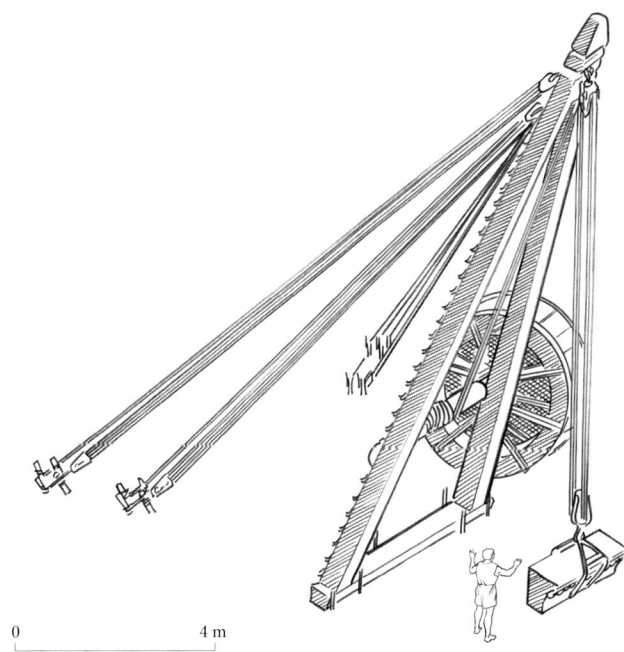

Reconstruction of ancient lifting machines based on text and surviving illustrations from On Architecture *by Vitruvuius.*

back to the capstan. Turning around the drum and axle, this rope winds itself up, so that the ends are stretched equally and the loads can be gently and safely raised. The capstan can be dispensed with, and the operation can proceed with greater speed, notes Vitruvius, "if a greater drum is fixed either in the middle or on one of the ends . . . and the drum, being trodden by men, can produce results more quickly."

The funerary relief of the builder Lucceius Peculiaris depicts the positioning of a column in the *scaenae frons* (stage background) of the theater at Capua (modern Santa Maria Capua Vetere), using a machine similar to that described by Vitruvius. In the image a crane is shown in conjunction with a large independent stepped wheel (Gr. *amphiesis*, Lat. *tympanum*), inside of which are two workers. Another important iconographic document for our understanding of ancient machines is the funerary relief of a member of the family of the Haterii, perhaps C. Haterius Tychicus, the contractor for several important construction sites during the reign of Domitian. The relief presents in detail a large hauling machine with seven beams, two in the front and five in the rear, each of which has a pulley. Attached to the machine is a stepped wheel, inside of which five workers are busy in its movement; on the outside another two men contribute to the effort by pulling on a cable tied to the spokes of the wheel, thus breaking its force of inertia.

Finally, Vitruvius (10.2.8–10) describes a third machine, "ingenious enough and suitable for speedy use," but cautions that "only skilled workmen can deal with it."

A pole is set up and is kept upright by cables in four different directions. Where the cables meet at the top, two sockets are fixed; the block is fixed to the sockets with ropes. Under the block is put a piece of timber about two feet long, six inches wide, and four inches thick. The blocks, with three sets of pulleys in their width, are fixed so that three guide ropes are inserted in the machine. These are brought down to the lower block and pass from the side next the pole over the upper pulleys; thence they are carried to the upper block

Above. Reconstruction of a model for centering a large span.

Opposite. View of the great arcade of the Aqua Marcia in Rome.

and pass over the lower pulley, from the outside to within.

When they come below they pass over the second pulleys from within outwards, and are brought back to the second pulleys above. Passing on they return below, and from below they return to the top. And passing over the top of the pulleys, they return to the lower part of the machine. Further, at the foot of the machine a third block is fixed; this is called *epagon* by the Greeks, *artemon* by us. The block is secured to its foot with three pulleys, over which the ropes pass, which are given to men to work. Thus three sets of men working without a capstan quickly draw a load to the top. This kind of machine is called *polyspaston* [a compound pulley], because with its many pulleys it is very easy and quick to work. The use of a single pole has this advantage, that by inclining it beforehand it can deposit the load sideways right or left as much as is desired.

Lifting a monumental element often posed a great challenge, rendering the successful outcome a genuine victory. A heroic participant in this stage of construction took pride in his accomplishment, seeing it as a feat of skill and strength. Two examples, although late in antiquity, are of importance in this regard: the Obelisk of Theodosius in the Hippodrome at Constantinople, the raising of which is depicted on the base of the obelisk itself; and the enormous monolithic stone vault of the Mausoleum of Theodoric at Ravenna, which still bears a series of projecting rings cut in the stone, through which the cables to lift it were passed.

CENTERING AND THE CONSTRUCTION OF ARCHES AND VAULTS

To construct an arch or a vault the first operation undertaken by builders was to erect centering, which constituted the support base for the structure to be built and reproduced an exact negative image of its outline. Centering was made mostly in wood, and composed of at least two arches joined crosswise by a curvilinear wooden assembly known as the substructure or mantle. The type of centering naturally varied according to the structure being built. It could rest directly on the ground with the help of a castle constructed of wooden poles, or it could be "flying." In the latter case, it was necessary to create putlog holes and brackets (Gr. *ota*, Lat. *ancones*) that corresponded to the uprights or support walls. These holes and brackets served for the insertion or support of the centering's crosswise boards or putlogs. On occasion the system involved the insertion of two pairs of brackets, one at the impost height of the arch or vault to be built and another at the height of the reins.

In more important constructions, such as the Porta Maggiore in Rome, and generally in triumphal arches, the corbels installed to support the centering were also designed to serve as decorative cornices. However, this was often the case even in relatively minor constructions, as indicated by some arches of the aqueduct at Divordurum (modern Metz) in France, and the even older arches of the Aqua Marcia in Rome. At Divordurum stone cornices were inserted in the imposts of the arches, isolated

Opposite. The enormous lacunar vaults of the Basilica of Maxentius in Rome.

inside a structure with facing walls made entirely in brick; for the arches of the Aqua Marcia a row of peperino blocks was located at the height that—together with another row along the keystones of the arches—formed a sort of stringcourse cornice inside the masonry in blocks of red Aniene tufa. In contrast, for the arches of the Aqua Claudia, the cornices were supported by pairs of brackets, located below and made of projecting blocks. Finally, for the construction of the arches of the Pont du Gard at Nîmes, whose span was more than 18 meters, an impost cornice, two series of putlog holes, and a continuous corbel at the reins were made in succession.

At major construction sites, efforts were always made to exploit the mechanical principle that governs arched structures, which consists in the fact that the area of the arch between impost and reins (within the section formed by an angle of about 30 degrees to each side) is, for all intents and purposes, stable because of friction. On this basis, for example, at the Baths of Caracalla in Rome, the vaulted roof of the large niches in the palaestrae were built up to the reins without the use of centering, by means of a progressive projection of the brick rows.

The construction of centering using containers of dirt was a distinctive procedure, as represented in the central cryptoporticus of the Domus Tiberiana on the Palatine, and in the smaller amphitheater at Pozzuoli (modern Puteoli). This system called for building vertical walls, by pouring concrete in deep reinforced trenches; containers of earth, shaped to correspond to the profile of the vault to be made, were then set up between the walls. Upon completion of the roof and the setting of the *opus caementicium*, workers began emptying the space by digging out the dirt. It was this method that formed the basis for the legend of the construction of the dome of the Pantheon. In the popular imagination the interior of the building had been filled with a pile of earth into which coins had been mixed; when the structure was completed everyone who helped to empty out the dirt had the right to keep any coins he found.

The Pantheon introduces another intriguing aspect of ancient building, the construction of vaults and domes in *opus caementicium*. Through experimentation in and development of construction techniques using this material, the Romans came to understand that its monolithic behavior was only theoretical. Thus, beginning in the Flavian period, the builders of vaulted structures began to insert parallel or crossing stiffening ribs made of brick, with the aim of controlling the mass of concrete and reinforcing the structure of the vault. This system, already present in the roofing of the ambulatory at the Colosseum, characterizes many buildings of Hadrian's Villa at Tibur (modern Tivoli), among others. The use of brick ribs was soon associated with an additional technical development, that of constructing a brick lining along the intrados surfaces of vaults, perhaps to improve the adherence of the plaster dressing.

Opus caementicium was not a light material, and the Romans found it necessary to resolve the problem of the great weight of vaulted structures, especially when they covered large spans. The difficulty concerned not just support walls but also coverings themselves, which were particularly stressed by their own loads. One of the methods used to reduce the mass of these structures was the insertion of terracotta amphorae in the conglomerate of the vaults, for these elements weighed far less than an equal volume of concrete. Examples of the use of this expedient are numerous, and include the Villa dei Gordiani on the Via Praenestina, the Temple of Minerva Medica, and the Mausoleum of Helena, which for this very reason is also known as the Tor Pignattara ("Tower of Vases").

A quite different method for addressing the substantial weight of *opus caementicium* was employed at the Pantheon. For the construction of the dome, which sat on a span

ENGINEERING AND TECHNIQUES AT THE WORK SITE 151

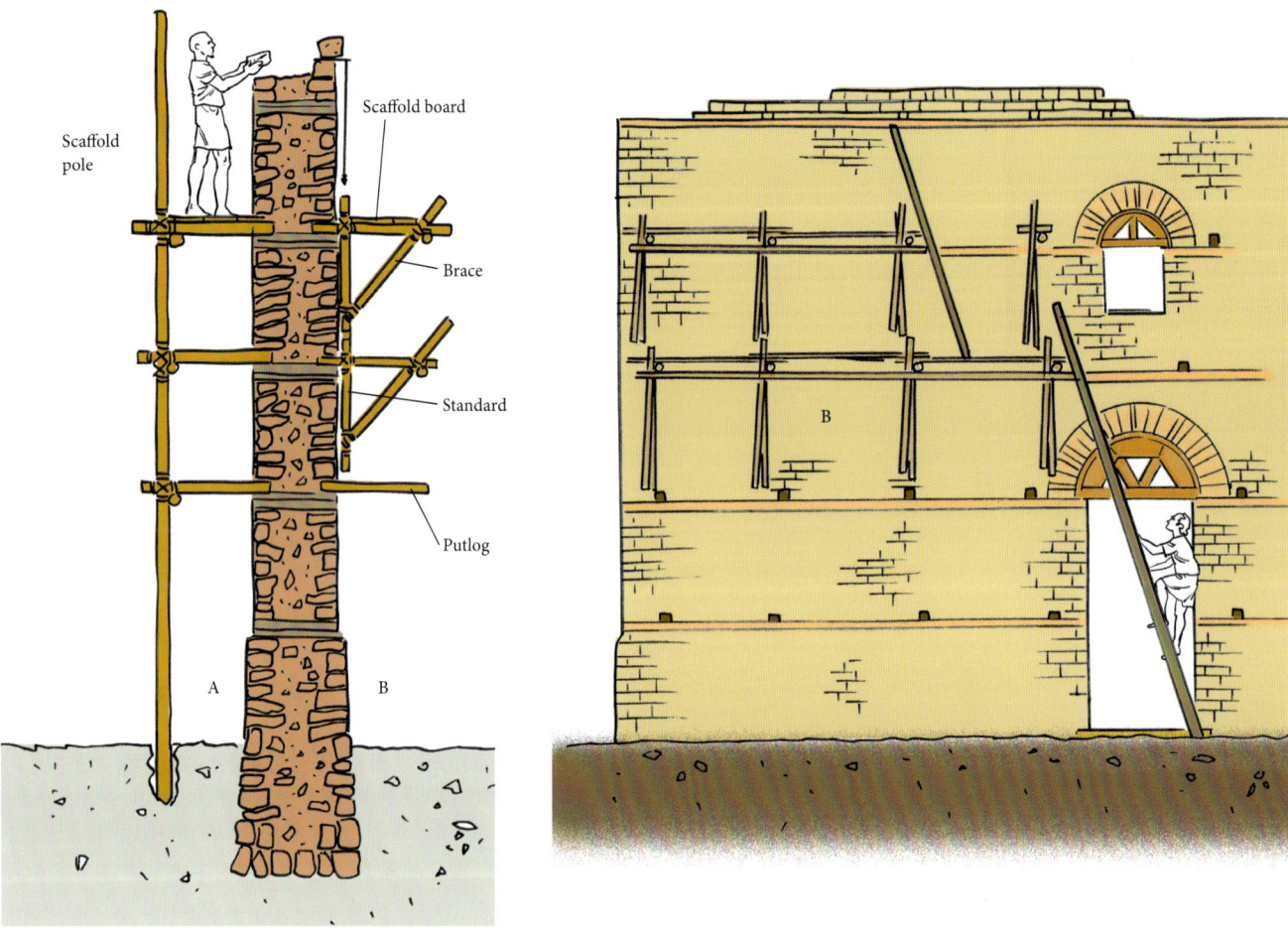

Two different types of scaffolding: A. Socketed; B. Cantilever.

of 43.3 meters, a concrete wall was composed of progressively lighter aggregates; shards of bricks were used in the area near the imposts, changing to fragments of tufa and broken bricks in the middle area, and ending in pieces of tufa and pumice stone near the great terminal opening. However, this strategy for the construction of the Pantheon did not constitute an isolated episode. Volcanic scoria had previously been used on the Palatine during the Domitian period and later for the Baths of Trajan, and was subsequently employed for the Baths of Caracalla and those of Diocletian. Outside Rome the use of pumice stone has been identified at Pompeii in Campania as well as at Capo Colonna in Calabria.

SCAFFOLDING

Properly functioning scaffolding, or staging (Gr. *ikria*, Lat. *machinae comparatae* or *scansoriae*), was essential to every ancient construction site. Workers and specialized artisans moved about on these structures, which were also used to store materials and tools. The basic component was a horizontal framework, or bridge (Gr. *xyloma*, Lat. *contignatio*), which could be connected by two simple trestles (Lat. *varae*), resulting in the most basic form of scaffolding, the mobile trestle. This type was used in the Roman period, as indicated by a stele (today in the museum of the Cathedral of St. Étienne at Sens in France, which depicts a scene of wall painting.

When the height of construction extended above three meters, it became necessary to erect scaffolding on several levels. It could be isolated and freestanding, or connected to the building; it could be erected inside the structure, outside it, or both. In each case the scaffolding was a wooden structure composed of standards, vertical poles fixed in the ground or set on a wooden base; putlogs, longitudinal pieces that ran parallel to the wall; and braces, shorter wooden elements affixed between the standards and putlogs.

Scaffolding was connected to a building in one of two ways. For a socketed system putlogs were fitted in openings, known as putlog holes, made in the walls. For cantilevered scaffolding poles were inserted in openings that extended the full width of the structure. This made it possible to erect, on the facade opposite the inner wall, a projecting scaffolding that rested on the extensions of the scaffold boards themselves. It was necessary to reinforce cantilevered scaffolding by inserting its projecting beams into a simple reticulated structure, composed of a vertical pole set against the wall, which functioned as a brace, and an oblique pole, which acted as a strut.

SUBSTRUCTURES

A substructure (Gr. *analemmata*, Lat. *substructiones*) is built above ground to level terrain and to contain the oblique thrust of an embankment (Gr. *choma*, Lat. *terrae congestio*). It could constitute a strong, continuous vertical wall or a system with internal or external buttresses (Gr. *antereismata*, Lat. *anterides*).

In the Archaic period the Greeks had often created broad terraces to locate their cult buildings on elevated sites. The Sanctuary of Hera at Argos is one of the oldest examples. In the seventh century B.C. a substructure in Cyclopean masonry was made there to contain the plateau that would support construction of the temple. But it was in the Hellenistic period that substructures acquired a truly monumental nature, thus presenting the opportunity to create massive terraced architectural compositions. Pergamum was the protagonist of this new style, and its architects are noted for their accomplishments. In the area of the Theater and Sanctuary of Demeter, the giant containment walls made in squared blocks and supported by long external buttresses were masterpieces of Greek engineering. Similar construction methods mark other great sanctuaries, including that of Asclepius at Kos and of Athena at Lindos, as well as Italic examples such as those at Praeneste (modern Palestrina) and Tibur (Tivoli).

Vitruvius (6.8.5–7) warns that care must be taken with substructures because "immense damage is caused by the earth piled against them." The change in seasons results in particular risks; a substructure "swells in the winter by absorbing water from the rains. Consequently . . . it bursts and thrusts out the retaining walls." To avoid damage, Vitruvius advises: "The thickness of the walling must answer to the amount of earth. Next, supporting walls or buttresses are to be carried up at the same time. The interval between them is to be the same as the height of the substructure, and the thickness that of the substructure. They are to project at the base in accordance with the thickness determined for the substructure. Then they are to be gradually diminished, so that at the top they may project as much as the thickness of the walling." Precautions must be taken on the inside of the substructure as well: "The wall must have projections like the teeth of a saw, such that the intervals between them are equal to the height of the substructure. The thickness of the teeth must be that of the main wall." Finally, the outer angles must be reinforced: "A diagonal wall is to be built between them, and from the middle of the diagonal wall, another wall is to be built to the interior angle of the main wall. The teeth and the diagonal walls will not allow the full pressure to fall upon the main wall but will distribute the thrust of the earth, which we have to hold in check."

ANCIENT HYDRAULICS: BETWEEN TECHNOLOGY AND ENGINEERING

"Water . . . by furnishing not only drink but all our infinite necessities, provides its grateful utility as a gracious gift."
Vitruvius, *On Architecture*, Book 8, Preface, 3

SPRINGS

Access to a source of potable water has always represented a fundamental prerequisite for the establishment of a city. As early as the Mycenaean period, large projects were undertaken in Greece to supply water for fortified citadels, and deep tunnels were often dug inside the circle of walls to ensure the continued availability of water in case of a siege. Mycenae was provided with a complex hydraulic system: water rose near the Lion Gate and from there was conducted along a channel that extended about 360 meters to a square cistern at the far northeastern end of the city, immediately outside the walls but in a naturally fortified site. To reach this basin, located roughly 18 meters below the level of the citadel, the Mycenaeans created a steeply descending passageway with a long staircase that was covered by a corbel vault, using progressively projecting blocks. During the same period similar installations were also made at Tiryns.

There were several springs along the slopes of the Acropolis at Athens. One of these was exploited in the Mycenaean period, although reaching it required excavation of a deep vertical gallery. History has recorded other springs that were progressively exploited for the collection and drawing of water. At the end of the sixth century B.C., a basin was built along the southern side of the Acropolis near the future Ionic stoa of the Sanctuary of Asclepius; the following century saw the monumentalizing of the Klepsydra spring, which rose along the northeastern slope. In both cases the spring was made sacred, and basins were built beneath ground level in contact with the bedrock to keep the water fresh and clean.

In many cases the water from a spring was directed into a channel, from which it flowed

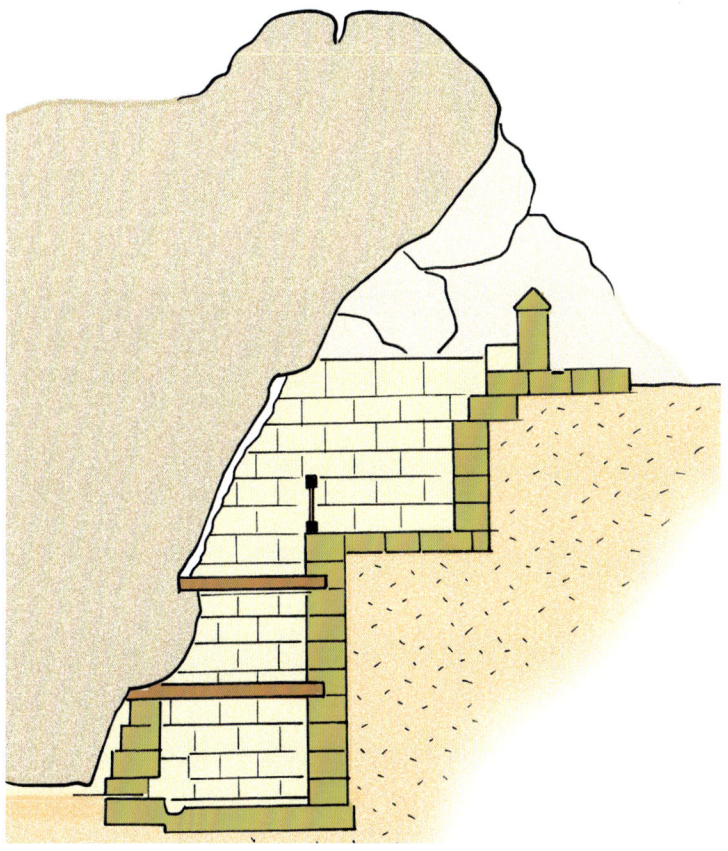

Opposite. Passageway with a false vault in stone blocks made at Mycenae to access water.

Bottom. Remains of the structures in limestone blocks of the Klepsydra spring in Athens.

ANCIENT HYDRAULICS: BETWEEN TECHNOLOGY AND ENGINEERING   155

View of the Tullianum with the modern altar commemorating saints Peter and Paul. Visible along the walls are the three courses of the original false dome constructed of stone blocks.

out of multiple spouts that were aligned above a single basin. This was probably the method used at the large, nine-spouted fountain house called the Enneakrounos ("Nine Springs"), built by the tyrant Peisistratus in the sixth century B.C. near the Kallirhoe spring in Athens, as well as at the fountain that was fed with water from the Kastalia springs at Delphi, dating to the same period.

Various springs were exploited in Rome, including one located on the eastern slopes of the Capitoline near the initial stretch of the street known as the Clivus Argentarius. In order to collect water the Romans constructed a circular cistern that was closed off from above by a sort of false dome built of peperino blocks. Named the Tullianum, this cistern was later transformed into a prison known in the Middle Ages as the Mamertine, traditionally believed to be the prison where the saints Peter and Paul were confined.

THE PEIRENE SPRING AT CORINTH AND ITS DRAINAGE CHANNELS

"Water will be more accessible if the springs flow in the open," writes Vitruvius (8.1.1), "but if they do not flow above ground, sources are to be sought and collected underground." The Greeks knew it was possible to reach the water-bearing stratum by digging long underground passages through rock. Sometimes, however, the hydrologic nature of the terrain made it impossible to identify and reach the site of the spring, and instead the tunnels functioned as large cisterns to collect water that dripped from their walls. These drain channels were dug so that the bottom sloped, causing water to flow toward an external basin.

The Corinthians were masters at building drain channels. Corinth alone had four structures of this type: three located in the lower city—the lower Peirene spring, the Glauke to the north of the agora, and the Lerna near the

View of the lower Peirene spring at Corinth. The oldest structures, incorporated into the arched facade from the Augustan age, were further monumentalized with the rearrangement of the complex in the second century A.D.

Sanctuary of Asclepius—and the upper Peirene spring on Acrocorinth.

The most monumental of these structures was the lower Peirene, located northeast of the access route to the agora via the Lechaion Road. Still active, the Peirene was named for one of the Naiads—daughters of the river god Asopos—who was carried off by Poseidon and bore him two sons, Leches and Cenchrias, for whom the city's ports were named.

In its first stage drainage for the lower Peirene spring was composed of four long, parallel channels that had been cut into the bedrock on which the eastern corner of the agora had been built. These emptied, each through five openings, into a tripartite basin that during the Hellenistic period was given an elegant facade with windows framed by Ionic pilasters. In front of this facade was a series of six deep, quadrangular basins from which it was possible to draw water.

The importance of water supplied by the Peirene spring explains various efforts at monumentalizing carried out on the complex by the Romans, beginning in the Augustan period with a new facade that included arches of stone blocks. Better documented were the transformations carried out in the second century A.D. by Herodes Atticus, who was also responsible for construction of the large nymphaeum at the Sanctuary of Zeus at Olympia. At the Peirene spring he gave the fountain a large courtyard with a central basin, framed by a rich structure with three exedrae dressed in polychrome marble.

WATER FROM WELLS

When springs were at a great distance, or when the water they supplied proved insufficient to meet a community's needs, the ancients dug wells. Far from a simple operation, the creation of a functioning well that

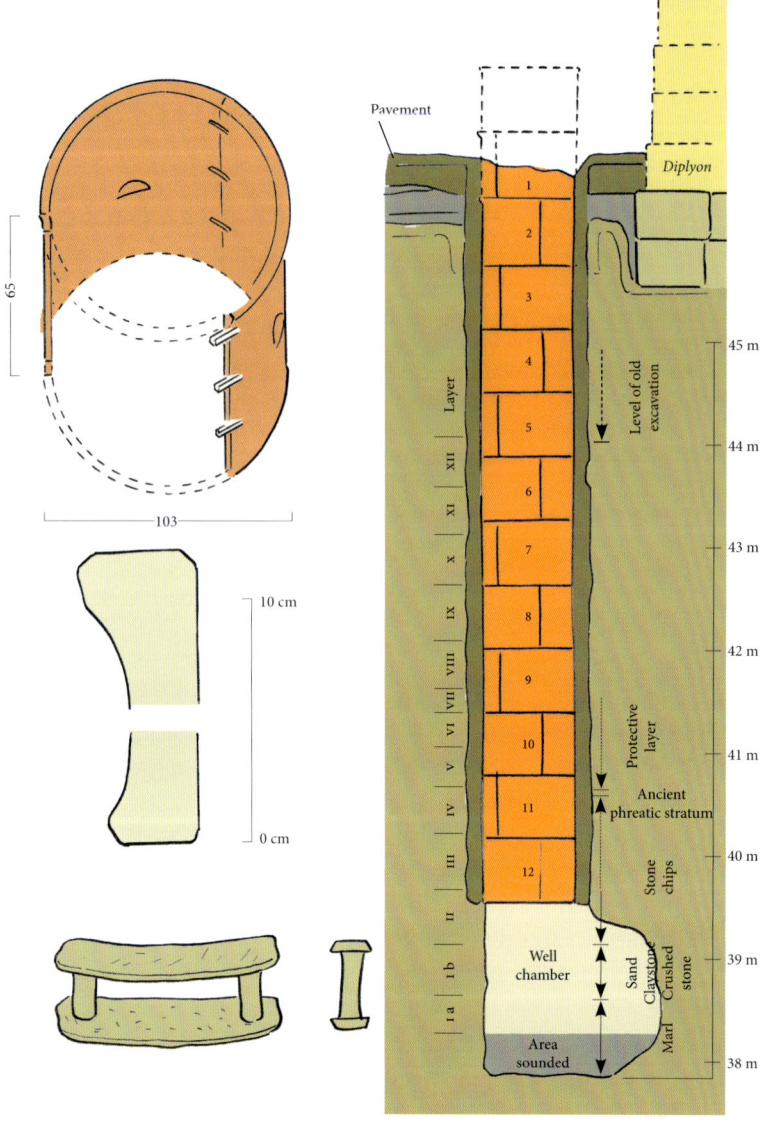

Cross section of one of the wells of the Dipylon at Athens, indicating the composition of the terracotta rings and the design of one of the cramps with which they were joined.

provided good-quality water was a true art, practiced by designated professionals (Gr. *phreorychoi*, Lat. *putearii*). "Those who look out for water must also observe the nature of the ground," advises Vitruvius (8.1.2), "for there are certain places where it rises. In clay, the supply is thin and scanty and near the surface; this will be not of the best flavor. In loose gravel the supply is scanty . . . this water will be muddy and unpleasant. In black earth, moisture and small drops are found; when these gather after the winter rains and settle in hard solid receptacles, they have an excellent flavor. But in gravel small and uncertain currents are found; these also are of unusual sweetness. In coarse gravel, common sand, and red sand, the supply is more certain, and this is of good flavor."

Once the nature of the terrain was identified, the next step was the actual digging, done by hand. During this stage, too, the work required skill, as the application of empirical experience did not always prove successful. Vitruvius described the excavation of exploratory shafts (8.1.4): "A hole is to be dug not less than three feet square and five feet deep, and about sunset a bronze or lead vessel, or a basin, is to be placed there. Whichever it is, must be smeared inside with oil and put upside down, and the top of the hole covered with rushes or leaves; and earth must be thrown above. On the next day it is to be opened, and if there are drops of water and moisture in the vessel, water will be found." He also cautions (8.6.12–13) that those who dig wells "must not make light of science," and points to the inherent dangers: "The methods of nature must be considered closely in the light of intelligence and experience, because the soil contains many various elements. . . . When these are heavy and come through the porous intervals of the soil to the wells which are being dug, they affect the excavators, in so far as the nature of the exhalation chokes the animal spirits in their nostrils. Hence those who fail to escape at once die there." Vitruvius advises the following precautions: "Let a lighted lamp be lowered. If it remains alight, the descent will be accomplished without danger. If, however, the light is extinguished by the power of the exhalation, then air shafts are to be dug right and left adjoining the well. In this way the vapors from the air will be dissipated, as the air is through the nostrils."

As digging slowly proceeded, the walls of the shaft were built up to seal them, as noted by Vitruvius (8.6.13): "When . . . we come to the water, then let it be enclosed by walling without blocking up the veins." Other systems

The great well of Kallichoron at the Sanctuary of Demeter at Eleusis. Note the accuracy with which the stones were dressed and their connection with metal cramps.

called for using a covering of wood panels, or large rings of terracotta fitted with foot holds to allow descent for inspection of the well.

Archaeological excavations in the agora at Athens have brought to light a number of wells, some of them dug as early as the second millennium B.C. The history of the area's changing use during the reign of Peisistratus has made it possible to attribute all the wells that were dug before the middle of the sixth century B.C. to houses and workshops. Subsequent wells, however, served the various civil and commercial activities that took place in this important sector of the city.

In the ancient imagination water assumed a purifying value and played a significant role in sacred rites. Consequently, wells were also commonly found at Greek and Roman sanctuaries, including the large Kallichoron well at the Sanctuary of Demeter at Eleusis, and the well located at the Sanctuary of the Chthonic Divinities at Agrigento.

UNDERGROUND AQUEDUCTS

Many ancient settlements arose near water sources, but in some cases the supply soon proved inadequate to the requirements of a growing population. Thus the need arose to construct aqueducts, even across long distances, to bring water to the city. Between the eighth and seventh centuries B.C. the Assyrian king Sennacherib had an aqueduct constructed at Nineveh, on the right bank of the Tigris, that was fully 55 kilometers long and delivered water from the Atrush and Kohan rivers.

Many ancient architects made their names solely on the basis of an aqueduct, which was a highly complex operation to design and build. Most important was the flow gradient of the channel along the aqueduct's route, since a slope was necessary to lead water in the proper direction. Modern measurements demonstrate that ancient aqueducts had a gradient that ranged between 0.0025 percent (along stretches of the Aqua Virgo in Rome) and 10

Right. Fountainhead of the aqueduct of Eupalinos on Samos.

Opposite, top. Double tunnel through rock of the aqueduct at Syracuse. Note the position of the nymphaeum on the right, above the Greek theater.

Opposite, bottom. The great nymphaeum dedicated to the Muses located at the end of the conduits of the Syracuse aqueduct.

percent (along portions of the Priene aqueduct in Asia Minor), with an average of 0.1 percent. Of course, it was impossible for workers to maintain a constant gradient, especially in the case of longer aqueducts that crossed a variety of terrains.

Sometimes an aqueduct did not draw its water directly from a spring or river, but began instead from a settling tank or fountainhead, which the water reached after traveling through smaller channels. Archaeologists have discovered both Greek and Roman structures of this type, such as the great fountainhead of the aqueduct designed by Eupalinos on Samos (sixth century B.C.), as well as that of the aqueduct built in the region of Eifel in A.D. 80,

160 ANCIENT HYDRAULICS: BETWEEN TECHNOLOGY AND ENGINEERING

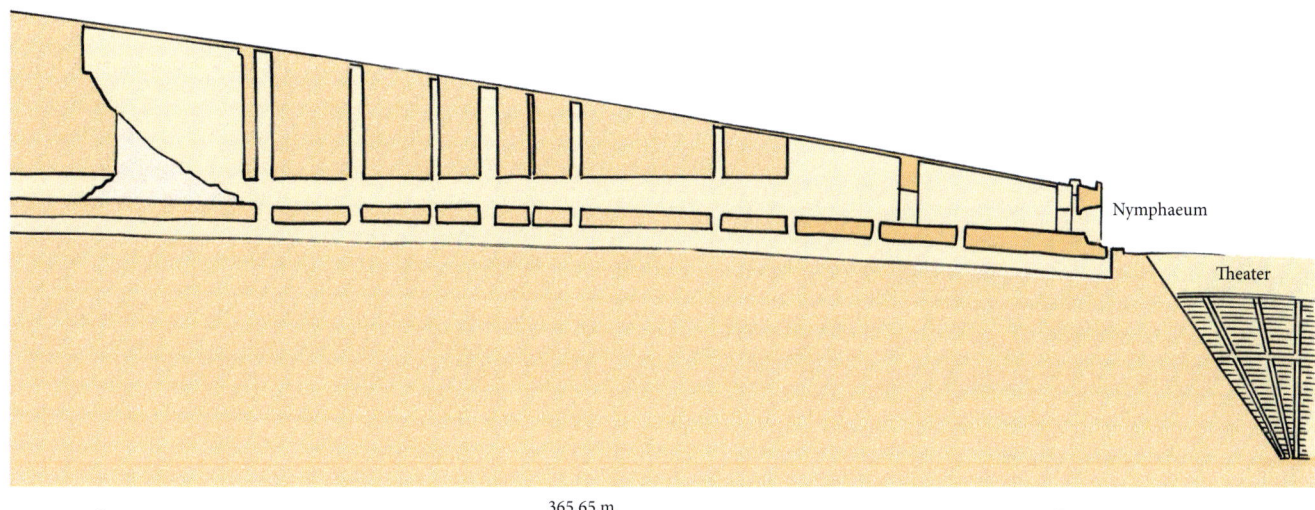

which traveled a route of about 95 kilometers and supplied water to the city of Cologne.

The greatest difficulty when building an aqueduct, as noted, was maintaining the necessary gradient for the flow of water; an excessive gradient could result in dangerous levels of pressure. Preserving the correct gradient was especially complicated when an aqueduct covered uneven ground or passed through an area with alternating valleys and hills. The methods to apply in such instances were outlined by Vitruvius (8.6.3): "If there are hills between the city and the fountainhead, we must proceed as follows. Tunnels are to be dug underground and leveled to the fall already described. If the formation of the earth is of tufa or stone, the channel may be cut in its own bed; but if it is of soil or sand the bed and the walls with the vaulting are to be constructed in the tunnel through which the water is to be brought." This process usually required the construction of tunnels wide enough to permit the movement of laborers who were digging. A series of vertical access shafts were opened along the length of the tunnel to facilitate the operation, and to provide later access for inspecting and cleaning the tunnel. Vitruvius (8.6.3) advised that "air shafts be at the distance of one *actus* [roughly 72 meters] apart."

ANCIENT HYDRAULICS: BETWEEN TECHNOLOGY AND ENGINEERING

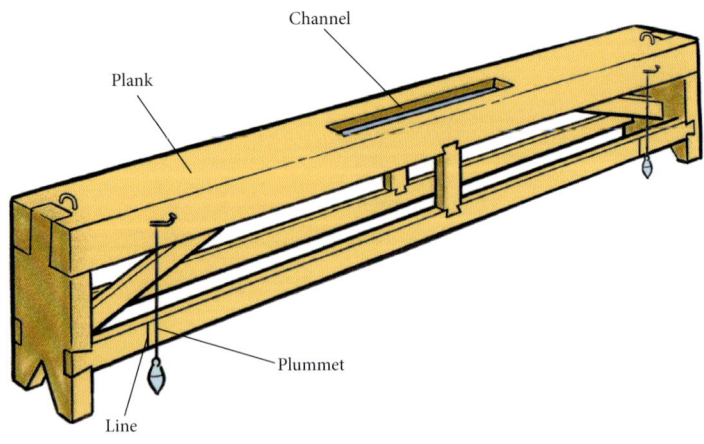

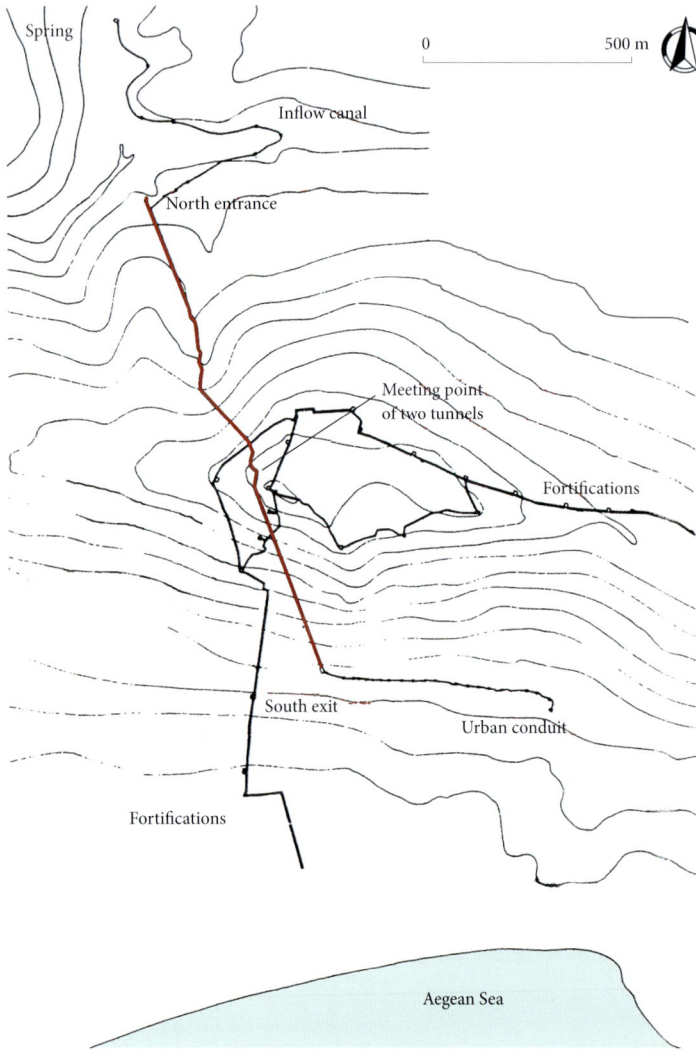

VITRUVIUS AND CALCULATION OF THE FLOW GRADIENT

As we have seen, Vitruvius's text constitutes an essential source of information on ancient building techniques. Addressing the means to supply water to country houses and towns, he notes (8.5.1) that "the first stage is to fix levels," a task accomplished "by *dioptrae* [a surveying instrument], or water levels," or when the *dioptrae* and the water levels are misleading, "by the *chorobates* . . . a straight plank about 20 feet long." Each end of the *chorobates* has legs that are fastened at right angles; and between the plank and the legs, cross pieces are joined by tenons: "These have lines accurately drawn to a perpendicular, and plummets hanging severally over the lines from the plank. When the plank is in position, the perpendiculars which touch equally and of like measure the lines marked, indicate the level position of the instrument." Noting the potential for wind to disturb the operation and make exact determination of a level position impossible, he specifies (8.5.2) that "a channel is to be put on the top side of the plank, five feet long, an inch wide, and an inch and a half deep. Let water be poured in. If the water evenly touches the lips of the channel, we shall know that the leveling is successful. Further, when we have leveled with the *chorobates*, we shall know the amount of the fall."

ENGINEERING AND HYDRAULICS IN GREECE DURING THE ARCHAIC PERIOD: EUPALINOS AND THE SAMOS TUNNEL

In the fifth century B.C. Herodotus (*Histories* 3.60) was the first to use the term *architect* in relation to Eupalinos of Megara:

> [The Samians] have three works greater than any others that have been made by Hellenes: first a passage beginning from below and open at both ends, dug through a mountain not less than a hundred and fifty fathoms [about 266 meters] in height: the length of the passage is seven furlongs [about 1,243 meters] and the height and

breadth each eight feet [about 2.4 meters], and throughout the whole of it another passage has been dug twenty cubits in depth and three feet in breadth [8.9 by 0.9 meters], through which the water is conducted and comes by the pipes of the city, brought from an abundant spring: and the designer [*architekton*] of this work was a Megarian, Eupalinos the son of Naustrophos.

Thus Eupalinos is credited with the creation of a great work of engineering on Samos, probably begun during the reign of the tyrant Aeaces around the middle of the sixth century B.C. and completed under the reign of his son Polycrates in the second half of the century. The tunnel's remains were identified in 1884. Measuring 1,036 meters long, it was dug into the rock of Mount Ampelos beginning at an altitude of 55 meters above sea level, and transported water that gushed from the Agiades spring, north of the city's walls. A walkway with an irregularly quadrangular section (roughly 1.8 by 1.8 meters) traveled alongside a deep, continuous trench that had been lined with terracotta to serve as a channel for the flowing water.

German archaeologists spent years excavating and studying the island's monuments to reconstruct the methods used to make the tunnel, and determined that it was created by digging simultaneously from both ends. Eupalinos applied great technical mastery to the resolution of the many problems connected to the tunnel's excavation, from his initial calculation of the location and direction of the two converging shafts to the transportation and disposal of materials unearthed, as well as maintenance of the channel's gradient at the level required to ensure the flow of water. As work advanced, from time to time it was necessary to alter—in various increments—the direction taken by the excavation, as indicated by numerous measurements marked in red on the rock walls by the ancient builders and still visible today. Cutting through the flank of the mountain, the tunnel passed beneath the city walls and therefore constituted a rapid and strategic means of passage to and from the city in the event of a siege. The tyrant Meandrius, for example, used the tunnel to flee the Persian attack led by Darius in 522 B.C.

The undertaking directed by Eupalinos on Samos was not without precedent. In Jerusalem, during the reign of Hezekiah (715–696 B.C.), a tunnel 537 meters long was constructed to bring water from the spring of Gihon to the Siloan Pool, located inside the city.

AQUEDUCTS UNDERGROUND AND ON ARCADES

Sextus Julius Frontinus was made commissioner of Rome's aqueducts (Lat. *curator aquarum*) in A.D. 97. He thus held one of the state's highest offices, responsible for control of the water supply to the city of Rome, including the administration of its aqueducts. Frontinus wrote a report on the city's aqueducts (*De*

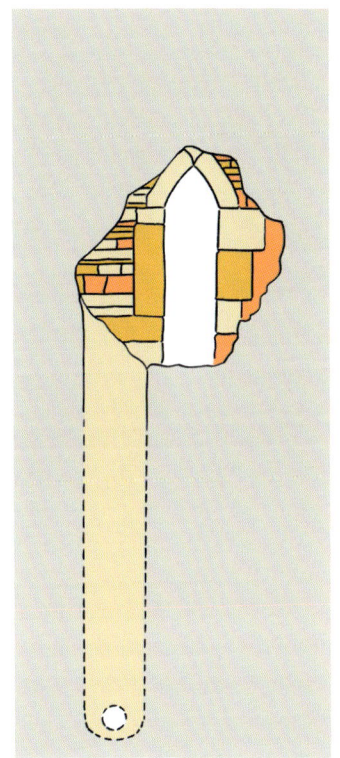
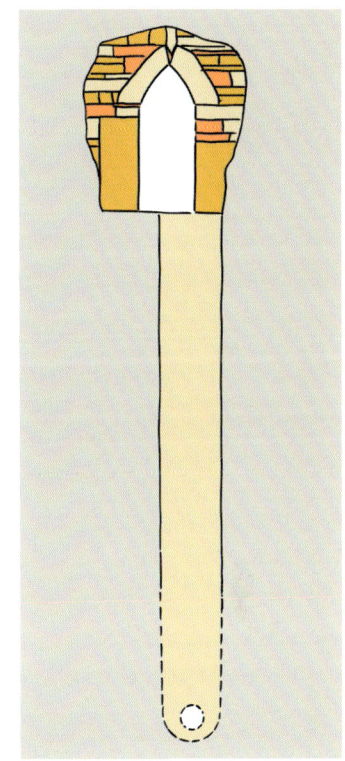

Opposite, top. Chorobates, as described by Vitruvius (8.5.1).

Opposite, bottom. Route of the aqueduct built by Eupalinos on Samos in the sixth century B.C.

Left. Two cross-section lengths of the tunnel of Eupalinos on Samos.

aquis urbis romae or *De aquae ductu*), in which he described three ways of conducting water: inside an underground channel (Lat. *subterraneus rivus*), through a tunnel (Lat. *cuniculus*), or along surface channels.

The system based on underground transport had been widely used by the Greeks, and was ancient even during Frontinus's time. It had the advantage of protecting water from external material and at the same time helped preserve the aqueduct itself. Underground structures of this type were often composite, with tubing, or channeling, carrying water within a larger canal made of stone slabs or terracotta. At Megara in Greece several stretches of an underground canal with tubing, dated to the Archaic period, have been found. Better preserved but dated later, to the fourth century B.C., are several covered conduits in Athens, made with walls and a covering of stone blocks; inside each canal the water runs through a terracotta channel with a U-section.

When the rate of water flow was sufficient, and the geological nature of the terrain made it possible, water was transported inside deep channels that had been dug in the bedrock. At Selinus several sections of the Greek aqueduct that began at a source in the town of Bigini reached depths of up to 6.4 meters. This technique was used even along rocky walls in instances when the decision had been made not to cut a tunnel through the rock. Traces of such channels can be seen at Syracuse on the vertical rock face of the Paradise Quarry. These channels were comparatively straightforward to dig, particularly when the rock face was soft; however, the segments were necessarily long, thus influencing the general gradient of the installation.

The Romans resolved this problem through the construction of aqueducts on arcades. In effect, these were bridges erected for the transport of water, and made it possible to maintain the necessary gradient despite the geological composition of the site. More importantly, the heights of these structures made it possible to reach even the most elevated urban areas. Aqueducts on arcades were constructed in all regions of the empire, and today they stand as one of the most characteristic monuments of the Roman period. Celebrated examples are to be found in Spain and France, such as the aqueduct at Segovia and the Pont du Gard at Nemausus (modern Nîmes), as well as those built by the Romans in Greece and Asia Minor, at Aspendos, Ephesus, and Nicopolis.

THE AQUEDUCTS OF ROME

"For four hundred and forty-one years from the foundation of the City, the Romans were satisfied with the use of such waters as they drew from the Tiber, from wells, or from springs. Esteem for springs still continues, and is observed with veneration." With these words Frontinus (*Aqueducts of Rome* 1.4) testifies that Rome was sufficiently supplied with water even before the construction of the great aqueducts. Water was drawn directly from the river, taken from various wells that had been dug in the city, or collected from a spring, such as that of the Tullianum.

The first aqueduct leading to Rome was the Aqua Appia. Its construction dates to 312 B.C., and it was ordered by the censor Appius Claudius Caecus, the same man responsible for the Via Appia (the Appian Way). The sources of the Aqua Appia were located in the area between the seventh and eighth milestones on the Via Praenestina. Constructed with a conduit roughly 16.5 kilometers long, most of the aqueduct was underground except for a brief section near the Porta Capena, between the Circus Maximus and the area of the future Baths of Caracalla. The canal was built in blocks of tufa, with a central channel, and its flow rate amounted to a daily supply of 73,000 cubic meters.

A few decades later, between 272 and 269 B.C., the censor Manius Curius Dentatus built a new aqueduct, the Anio, later called the Anio Vetus. With a length of about 63 kilometers, it drew water from the Aniene River and brought

The aqueduct on arches at Nicopolis in Epirus.

it up to the northeastern quarters of the city of Rome. In this case also the aqueduct was located in large part underground, with some sections built in small blocks of tufa that were covered with slabs of cappellaccio. The Anio Vetus was the first aqueduct to reach the city at the site known as "ad Spem Veterem" ("to the Ancient Hope"), in the eastern area of the Esquiline at the point where the Porta Maggiore was later to stand. The daily supply of this aqueduct was 176,000 cubic meters.

In 144 B.C. the praetor Quintus Marcius Rex ordered the building of the Aqua Marcia, which is still active. The aqueduct originated in the springs of the Aniene River and extended for a length of about 91 kilometers. It entered the city, like the Anio Vetus, in the area of Porta Maggiore and extended north, past the Porta Tiburtina; various city conduits branched off the Aqua Marcia, one of which was the first aqueduct to reach the Capitoline. Much of the Aqua Marcia remained underground, and as Pliny the Elder recalls (*N.H.* 36.121), it had been made by driving "underground passages through the mountains"; some sections were carried on arches that are still partially preserved along the Via Latina. The Aqua Marcia was capable of supplying Rome with 187,600 cubic meters daily.

The demand for water in Rome continued to grow, and by 125 B.C. another aqueduct, the Aqua Tepula ("Tepid Water"), was constructed, with its origin in the area of the Alban Hills near the town of Marino. Named for the lukewarm temperature of its water, this was a comparatively short conduit, only about 18 kilometers long, and capable of bringing only 17,800 cubic meters of water to Rome daily.

In the first century B.C. the emperor Augustus ordered that another three aqueducts be built. The first, the Aqua Iulia, was constructed in 33 B.C. by the Roman general Agrippa and originated in the same area as the Aqua Tepula. The second aqueduct, far more

The branch of the Aqua Claudia built to supply the imperial palace on the Palatine, as represented in a model at the Museo della Civiltà Romana, Rome.

Opposite. Section of the Aqua Claudia.

powerful, was the Aqua Virgo, named for the young girl who had indicated the site of the spring to soldiers assigned to the search for water. Built by Agrippa in 19 B.C., it reached the city on the slopes of the Pincian Hill, continuing on arches to the Campus Martius as far as the area of the Pantheon and the Baths of Agrippa. Still active today, the Aqua Virgo feeds the well-known Trevi Fountain. The third aqueduct ordered by Augustus was the Aqua Alsietina, built in 2 B.C. The three conduits differed greatly in the amount of water supplied: the largest was the Aqua Virgo, which delivered 100,160 cubic meters a day; followed by the Aqua Iulia, which transported 48,240 cubic meters; and finally the Alsietina, which brought only 15,680.

Rome received another two aqueducts during the reign of Claudius: the Aqua Claudia and the Anio Novus. The construction of both had begun in 38 B.C., during the reign of his predecessor Caligula, but they were not completed until A.D. 52. Construction of the Aqua Claudia was an undertaking of enormous proportions. The conduit was more than 68 kilometers from end to end, and extended along tufa arcades to reach the area of "ad Spem Veterem" at the point where the emperor Claudius built a double monumental arch, the Porta Maggiore. From this point the emperor Nero subsequently ordered a branch built to supply water to the buildings on the Caelian Hill, and Domitian subsequently extended this conduit to reach the Palatine. The daily supplies of the Aqua Claudia and the Anio Novus were 184,280 and 189,520 cubic meters, respectively.

The last two aqueducts constructed in Rome were the Aqua Traiana, built by Trajan in A.D. 119 to transport water from springs near Lake Bracciano, and the Aqua Alexandrina, made around A.D. 226 by the later Roman emperor Alexander Severus. Since these final two aqueducts were built after Frontinus wrote his text, we have no source to inform us of their daily output. Even without the Aqua Traiana and the Aqua Alexandrina, however, the nine older aqueducts alone brought Rome a daily water supply equal to 992,200 cubic meters.

CONSERVATION OF WATER: CISTERNS AND RESERVOIRS

Since earliest times humans have recognized the need to collect and save rainwater. In the Minoan period on Crete cisterns were made to hold rainwater, as indicated by those still well preserved in the palace at Zakros.

Vitruvius (8.2.1) claims that rainwater "has more wholesome qualities, because it comes from the lightest and most finely tenuous of all sources; then filtering through moving air, it liquefies in storms and so reaches the earth." Earlier the Greeks had seen rainwater as "sent by Zeus," and its importance was such that Aristotle (*Politics* 1330b) advised every city to have "a plentiful natural supply of pools and springs, but failing this . . . an abundance of large reservoirs for rainwater, so that a supply may never fail the citizens."

In both the Greek and Roman worlds, efforts were consistently made to collect rainwater in cisterns so that it could be reserved and made available for future use. Numerous

Top. The great arched cisterns built in the Hellenistic period near the theater of Delos.

Opposite, top. Reconstruction of the water-purification system in the castellum aquae at Pompeii.

Opposite, bottom. Base of one of the columns in the Yerebatan Sarayi at Istanbul.

ancient cisterns, both public and private, can be found on Delos, Santorini, and other Cycladic islands, and demonstrate that the shape of Greek peristyle houses, like the shape of Roman houses with an atrium, was at least partially related to a desire to collect rainwater in cisterns designed exclusively for that purpose, usually located beneath the floor of an internal court. From there water could be drawn off by way of special wells, most often fitted with a wellhead, which was ornamented with varying degrees of decorations. The transition from the Roman period to late antiquity saw the creation of gigantic open-air reservoirs such as that at Silifke in Cilicia and, even more massive, the cisterns of Aspar and Aetius at Constantinople, of such considerable size that today they are used as sports centers.

Reservoirs made to hold water transported by aqueducts or through channels differ from those described above. These structures, which are also often quite large, were usually built at the termination of a major water channel, where they frequently supplied fountains and nymphaea. All the larger bathing complexes had water reserves of this kind. The Baths of Caracalla, for example, were provided with water by a branch of the Aqua Marcia and contained a water tank that could hold more than 80,000 cubic meters. Istanbul has a perfectly preserved ancient water cistern, the Yerebatan Sarayi ("Sunken Palace") cistern, a gigantic structure built by Constantine IV in the fourth century A.D. Fed by an aqueduct initiated under Hadrian and continued under Valens, it measures 140 by 70 meters and is divided into corridors by 336 marble columns taken from other structures.

DISTRIBUTION OF WATER: CASTELLUM AQUAE, CANALS, AND PIPES

Pompeii represents a privileged site in terms of our understanding of the methods of water distribution at a Roman center, as the eruption in A.D. 79 completely buried the ancient city, preserving its structures and operating systems. Water arrived at Pompeii at the highest part of the urban area, near the Vesuvius Gate, thereby facilitating the transport of water through pipes that led to individual city neighborhoods. The Vesuvius Gate was the site of the *castellum aquae*—a building with a trapezoidal layout and a circular space covered by a dome (5.7 meters in diameter)—which functioned as a purification and distribution tank. As water from an exterior conduit entered the first of two tanks, it passed through a grid that served as the first filter; a second grid, with a finer mesh, was located at the center of the structure. Inside the basin water was purified by natural decantation, and only the uppermost, clean water spilled over into the second tank, through a lead pipe. At this point the water was ready for distribution to the various city quarters.

Water pipes were generally made from either terracotta or lead. Those in terracotta, more ancient in origin, consisted of tubular elements modeled in a slightly conical shape so that they could be joined using a "male–female" insertion system. For this reason every length of pipe had a slightly projecting strip around each end that served as a sealing ring. Some of the plainer versions of pipe were made with a narrowed neck, but these did not function as efficiently. Once the elements were joined, they were sealed with an adhesive mortar. "Earthenware pipes are to be made not less than two inches thick," writes Vitruvius (8.6.8), "and so tongued that they may enter into and fit one another. The joints are to be coated with quicklime worked up with oil."

While certainly more expensive, lead pipes offered numerous advantages in terms of the seal they provided and their strength, and because they could be specially calibrated. The Romans, in fact, had normalized the quantity of water that passed through pipes of standardized diameters, measured in quarters of a digit (*quadrantes*) and digits (*digiti*). The unit of measurement for water was the *quinaria* (4.2 square centimeters), corresponding to the flow rate of an element with a diameter of 5 *quadrantes* (23 millimeters). Pipes were made by bending sheets of lead around a calibrated cylinder, then soldering the edges with molten lead. Unlike clay pipes, those made of lead were joined by inserting one inside another; the two pipes were secured with a short coupling, and the three elements were soldered together with molten lead. Water pipes were often given a seal indicating the owner, the maker, the building to which the pipe belonged, and even the emperor.

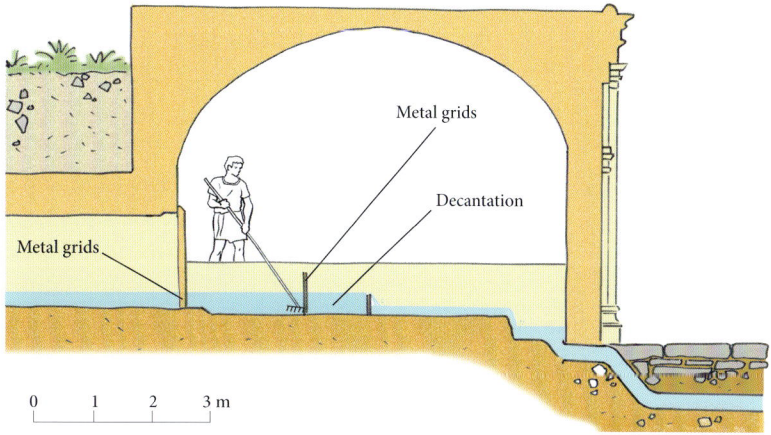

Bottom. Large lead water tube, with impressed seal, found at Verona.

Opposite. Remains of a large architectural fountain, with an Ionic facade distyle in antis, *built near the theater at Ephesus.*

FOUNTAINS AND NYMPHAEA

Recognizing that water is necessary to all human activities, as early as the sixth century B.C. the principal cities of the ancient world had built fountains for the public distribution of water. As noted, in Athens Peisistratus and his sons constructed the nine-spouted Enneakrounos, which is perhaps identifiable with the structure discovered on the southeastern corner of the agora. During the imperial period the agora came to have at least five monumental fountains.

Within sanctuaries fountains were of a different kind, since water played an active role in the rituals performed in the sacred area. In the cults of health-related divinities, such as Apollo and Asclepius, this aspect of water was further accentuated by the many healing rituals in which it was involved. In fact, water from fountains not only was drunk but also was used for purification, either by partial ablution or total immersion.

The ornamental aspects of water were emphasized in the Roman period. Water made its presence known in homes, villas, and gardens by way of decorative fountains embellished with jets and other displays of water, and in cities by the construction of large public fountains. The most famous such fountain in Rome was the Meta Sudans ("Sweating Marker"), from the Domitian period, a tall conical structure designed so that water slid down its sides as though "sweating."

Far more monumental were the nymphaea of the imperial period. These structures, named for water nymphs, were often composed of several levels and embellished with sculptures and rich decoration in polychrome marble. Some nymphaea, such as those in the cities of Ephesus, Miletus, Perga, and Side in Asia Minor, resembled large theatrical scenes (Lat. *scaenae frontes*), and the water splashing out from different points was collected in a large lower basin. Other nymphaea, such as those in Athens and Gortyna on Crete, were composed of a large semicircular or rectangular exedra, from which opened numerous niches decorated with statues.

LAVATORIES

One of the architectural typologies that the Romans spread throughout their empire, beginning in the first century A.D., was the lavatory (Lat. *foricae*). The presence in Rome of public latrines from as early as the second century B.C. testifies to differences between the Roman attitude in this regard and the earlier, more discreet view of the Greeks. Indeed, it seems that the Romans felt no shame concerning their physiological needs and shared the moments spent in latrines with others, conversing with anyone who happened to be present.

170 ANCIENT HYDRAULICS: BETWEEN TECHNOLOGY AND ENGINEERING

Opposite. The monumental latrine at Hierapolis in Phrygia. Clearly visible is the large outer channel over which a long row of seats, today lost, was originally positioned.

Aside from layout and decoration, the operation of every latrine was based on the same, relatively simple system. A continuous bench (Lat. *sella*), usually set on top of a raised platform, ran along one or more walls of the latrine; a regularly spaced series of circular openings appeared along the bench, with each aperture extending down and toward the front of the seat. Below the openings ran a canal with a steady stream of water that flowed to the nearest sewer; at the base of the bench ran another stream of water. It seems that handled sponges used for personal hygiene were dipped into this water; both the philosopher Seneca (*Epistles* 70.20) and the Latin poet Martial (*Epigrams* 12.48.7) make reference to them.

Latrines were usually located in heavily trafficked areas. In Athens a construction of this type was made at the southwestern corner of the agora, and another immediately outside the great square that served as the agora under Roman rule. At Ostia a public latrine was built immediately to the south of the forum, beside a large nymphaeum. In other cases latrines were placed within larger architectural complexes, such as baths. At Ephesus the large latrine connected to the Baths of Scolastica, along Curetes Street, still preserves a good portion of its seats and canals. Again at Ostia, in the latrines near the Forum Baths, both the seats and the canal beneath them were made of white marble.

THE CLOACA MAXIMA: A GREAT HYDRAULIC SYSTEM IN THE HEART OF ROME

The largest city of the ancient world arose and grew in an area whose famous seven hills (Lat. *septimontium*) articulated the surface through a series of slopes and rolling valleys in such a way that rainwater, along with water from the area's various springs, flowed into the Tiber by way of several natural streams. Ancient sources refer to the Amnis Petronia, which collected water from the slopes of the Pincian and the Quirinal. There was also the Spinon, fed by waters converging between the Quirinal and the Esquiline. Even so, some of the areas of Rome destined to be of central importance to the future rise of the city were frequently swampy, and artificial sewers (Lat. *cloacae*) were built there as early as the end of the seventh or early sixth century B.C.

Pliny the Elder (*N.H.* 36.105–6) attributed the beginning of these constructions to Tarquinius Priscus, the fifth king of Rome, and described the system of great conduits in the capital city during the early imperial period:

The city sewers [are] the most noteworthy achievement of all, seeing that hills were tunneled and Rome . . . became a "hanging" city, beneath which men traveled in boats during Marcus Agrippa's term as aedile. . . . Through the city there flow seven rivers meeting in one channel. These, rushing downward like mountain torrents, are constrained to sweep away and remove everything in their path, and when they are thrust forward by an additional volume of rainwater they batter the bottom and sides of the sewers. Sometimes the backwash of the Tiber floods the sewers and makes its way along them upstream. Then the raging flood waters meet head on within the sewers, and even so the unyielding strength of the fabric resists the strain. In the streets above, massive blocks of stone are dragged along, and yet the tunnels do not cave in. They are pounded by falling buildings, which collapse of their own accord or are brought crashing to the ground by fire. The ground is shaken by earth tremors; but in spite of all, for 700 years from the time of Tarquinius Priscus, the channels have remained well-nigh impregnable.

Livy (*History of Rome* 1.38.6) explains that sewers were needed to drain the swampy zones between the city's hills as well as the area around the Forum, since these "were too flat to carry off the flood waters easily." He attributes construction of the Cloaca Maxima to the

last king of Rome, Tarquinius Superbus, and explicitly notes (1.56.2) that the sewer was "a receptacle for all the offscourings of the city." Amazingly, this conduit has operated for more than two and a half millenia, serving not only the ancient city but also that of the Middle Ages and the modern era.

The contorted route of the Cloaca Maxima followed the course of the natural stream that ran from the Subura quarter, between the slopes of the Quirinal, Viminal, and Esquiline hills, to empty into the Tiber near the Ponte Rotto (the ancient Pons Aemilius). The sewer then crossed the areas of the imperial forums—passing beneath the Forum of Nerva—and the Forum Romanum, and entered the area of low ground known as the Velabrum. The conduit of the Cloaca Maxima was a long barrel-vaulted tunnel that in its largest sections reached a height of 3.3 meters and a width of 4.5 meters. Some scholars believe that the sewer originally lacked a roof—its contents flowing open to the air—but was crossed by a series of small wooden bridges, as suggested by holes that remain in the tunnel and would have held the beams; it cannot be ruled out, however, that those openings served instead as putlog holes for the scaffolding used to construct the tunnel's vaults. The building methods and construction materials identified along the long route of the sewer testify to the presence of numerous undertakings and reconstructions that took place during the republican period and the empire. Some of these can be related to episodes mentioned in literary sources; for example, there were the measures taken by Agrippa, who, according to Dio Cassius (*Roman History* 49.43), "cleaned out the sewers" in 33 B.C.

A section of the Cloaca Maxima, still in use as a sewer, is located below the colonnade (Lat. *porticus*) behind the Forum of Nerva. Today it runs at a depth of about 12 meters, but in ancient times its bed was roughly 5 to 6 meters below street level. The structure, dated to the Augustan period, is a vaulted tunnel about 3 meters wide and 3 meters high, built entirely in blocks of red Aniene tufa. The very next portion of the sewer, near the Temple of Minerva, displays a different construction system, with a vault made of blocks of peperino and walls in *opus caementicium*, which were given a waterproof seal with a thick layer of ground pottery.

Farther on, the sewer reached the area of the Forum Romanum, passing below the Basilica Aemilia with a conduit built between 55 and 34 B.C. in blocks of travertine and Aniene tufa. Near the basilica's steps, at the point where the sewer entered the area of the forum, stood the small but important shrine of Venus Cloacina, a divinity who protected the good functioning of the Cloaca Maxima and the entire sewer system of Rome.

After receiving water from the sewer that ran beneath the Via Sacra, the conduit obliquely crossed the forum. Archaeological research has revealed that, during the Late Republican period, the Romans resolved the difference between the height of the sewer and the low level of the forum's floor by dividing the conduit into a double tunnel built in *opus incertum* and *reticulatum*. The eastern section of this system has been completely excavated and today can be visited.

Leaving the forum and passing below the Basilica Julia, the Cloaca Maxima continued toward the Tiber, following the route of the ancient street called the Vicus Tuscus ("Etruscan Street"). Here the canal, made entirely in *opus caementicium* mixed with flint, narrowed to a width of only 1.6 meters. Probably built in the early Augustan period, this segment of the Cloaca Maxima replaced a more ancient canal, which was distinguished by a sloping roof and the use of blocks of cappellaccio.

Having passed the Arch of Janus, the sewer reached the Forum Boarium and beyond it the Tiber, into which it emptied at a point opposite the ancient road that preceded today's Lungotevere. The conduit ended with a triple archivolt arch, constructed entirely in blocks of gabina stone and inserted in a wall made with squared blocks of Grotta Oscura tufa.

HEATING SYSTEMS AND BATHS

BATHS OF THE GREEKS

As early as the Minoan period, the Greeks attributed a sacred meaning to water and instituted complex rituals related to it, as indicated by the numerous hydraulic installations found in various Cretan palaces as well as by the spaces made expressly for "baths," such as the so-called Queen's Bath in the palace of Knossos. In addition, there were numerous basins and fountains with purifying waters, usually composed of a large basin built into the ground and including a stairway access. Several structures of this sort, rectangular and circular in form, can be found in the palace of Zakros on the far eastern end of the island

It is true, however, that during the historical period the Greeks did not demonstrate as great an interest in water as did the Romans, and limited themselves to baths for cleansing and rituals, either partial ablutions or full-body baths involving immersion in cold water. Only in the Hellenistic period did spaces reserved for baths spread to private homes. In the known examples, bathing involved tubs (Gr. *pyeloi*) made of terracotta and somewhat small in size (1 by 0.5 meter on average); some of these have the characteristic shape of a modern sitz bath, with a small seat in the rear and a small circular cavity on the bottom.

More common, and dating back to the Archaic period, was the custom for Greek athletes to take a bath at the end of physical exercise, and an area was set aside for this activity in all gymnasiums. In its most ancient phase bathing was extremely simple and accomplished at a circular basin (Gr. *louterion*), usually set on top of a pedestal. Only near the end of the fifth century B.C. were special sites (Gr. *loutra*) built for bathing, with rectangular tubs arranged in series along one or more walls of a room. Water often arrived in the tub by way of down-flow holes located above it, sometimes running from one tub to the next through short connection canals. The gymnasium built at Delphi in the fourth century B.C., located just west of the Sanctuary of Athena Pronaia, preserves several portions of a room made for the

The monumental fountain in the Minoan palace at Zakros on Crete.

Tubs in the bathing facility at the Sanctuary of Zeus at Nemea.

athletes to bathe, and includes a large circular basin and a row of tubs along the terraced wall of the upper open-air track (Gr. *paradromis*). Similar installations are preserved in the gymnasiums of Eretria and Amphipolis. Also during the fourth century B.C., at the Sanctuary of Zeus at Nemea—which in 573 B.C. had become the site of the Pan-Hellenic games—a large two-story guesthouse (Gr. *xenon*) was made to house the athletes. A large bath, divided in two distinct spaces, was built adjacent to this structure near the end of the century. The first area was partitioned into three sections by means of two rows of columns; in the middle of the central area was a low pool, while each of the side areas had four stone tubs for individual bathing. The second space, with a central row of four columns, served all the secondary activities related to bathing. The entire complex was fed with water from a spring located a short distance away.

During this same period there were also several bathing establishments for the public (Gr. *balaneia*), but they were relatively small installations and usually organized around a main room, circular or rectangular, with tubs aligned along its walls. Structures of this type can be found in Athens and Eleusis and on Delos, and in the West at Velia and Megara Hyblaea, as well as at other places. Generally these were cold baths; Plato (*Laws* 6.761c) suggests that warm baths were best for the old, the sick, and "those whose bodies are worn with the toils of husbandry." For these individuals water was warmed in metal containers set on top of braziers.

THE BATH IN THE SANCTUARY OF ASCLEPIUS AT GORTYS

The remains of one of the first examples of a room heated by means of tubes passing beneath the floor can be found at Gortys in Arcadia. This space was a bath inserted in the important Sanctuary of Asclepius, and related to the presence of a spring of water with salubrious properties. Beginning in the fourth century B.C., the

View of the bath complex at the Sanctuary of Asclepius at Gortys in Arcadia.

entire area had been monumentalized by the creation of a large peripteral hexastyle temple (23.6 by 13.2 meters), for which the celebrated sculptor Scopas made the statues of Asclepius as well as the goddess of health, Hygieia, as noted by the Greek traveler and geographer Pausanias (*Desc. of Greece* 8.28.1).

A bathing complex was also built at Gortys in the fourth century, immediately to the south of the temple building. It may originally have been intended as a structure to lodge pilgrims (Gr. *katagogion*), but was soon transformed by the insertion of a central circular room and other spaces related to the practices associated with healing baths.

Access to the structure was by way of a rectangular entrance hall decorated with semicolumns and connected to a vestibule with a large semicircular niche. From there one reached a large circular hall with two semicircular niches; in this hall were several benches and pools for preliminary ablutions. To the north was another circular space with a continuous series of vertical tubs similar to sitz baths along its walls; to the northeast was yet another circular room. Some of these rooms were heated by way of warm air that passed beneath the flooring, similar to the system that later was the basis of Roman baths. Indeed, the last room, which was heated but had no tubs, was probably designed for steam bathing.

A large, square water cistern was located along the western side of the complex. This was paved in pieces of terracotta, and its walls were covered with a good-quality hydraulic mortar, spread so as to create along the inner corners a convex seam that would facilitate cleaning.

BATHS OF THE ROMANS

It was during the third century B.C. that public baths began to take hold in Rome. Many Romans followed the tradition of bathing at home, usually inside a small room (Lat. *lavatrina*) that was located next to the kitchen to profit from its heat. In the transition from the end of the Republican period to the beginning

The mosaic in the Baths of Neptune at Ostia.

of the Imperial, there was heated debate in Rome between those who remained anchored to the austerity of the past and those who sought instead the luxury and physical pleasures of the new bathing establishments.

Pliny the Elder (*Natural History* 36.121) recalled that "Agrippa, moreover, as aedile [33 B.C.] added to these [older aqueducts] the Aqua Virgo, repaired the channels of the others and put them in order, and constructed 700 basins, not to speak of 500 fountains and 130 distribution reservoirs, many of the latter being richly decorated. . . . He himself in the memoirs of his aedileship adds that in celebration games lasting for fifty-nine days were held, and the bathing establishments were thrown open to the public free of charge, all 170 of them." Subsequently, in 25 B.C. Agrippa, as a private individual, built a large bathing complex at the Campus Martius, granting it to the people of Rome with free admission. Sadly, very little of this complex remains, but it is known from several Renaissance drawings as well as a fragment of the *Forma Urbis Romae*, a map dating to the Severan period. The structure was laid out around a large circular hall with a diameter of 25 meters, still partially preserved.

Although bathing installations continued to be built in Rome, Seneca (*Epistles* 86) reports that in the Middle and Late Republican periods these public baths were not yet particularly luxurious, as the water did not run through pipes and the rooms were simply plastered. Only under Nero and then the Flavians was the city given bathing establishments on a monumental scale, enriched by palaestrae, pools, and large heated rooms. At the same time cities of the empire began building lavish bathing complexes, decorating them with statues and colored marble.

Rome was without doubt the ancient city with the greatest number of bathing installations. A sense of their widespread presence in the middle of the fourth century can be gleaned from the catalogues of Roman landmarks known as the Regionaries, which indicate that Rome then had between 856 and 951 individual baths along with eleven large bathing complexes.

PUBLIC BATHS AND THE COURSE OF A VISIT

Although Vitruvius lived at a time before the spread of large bathing complexes, he does refer (5.10.1–5) to the construction of baths, and discusses the best orientation for a structure to ensure that it can exploit the warmth of the sun: "Firstly a site must be chosen as warm as possible, that is, turned away from the north and east. Now the hot and tepid baths are to be lighted from the winter west; but if the nature of the site prevents, at any rate from the south. For the time of bathing is fixed between midday and evening."

Roman baths were organized as a series of rooms with water at different temperatures. The coldest was the *frigidarium*, where the water remained at room temperature; in the *tepidarium* water was kept lukewarm; and the *caldarium* contained warm water. Although each of these spaces could be articulated in a variety of ways and more or less decorated, their functions did not vary from site to site. What altered was the composition of other spaces in the bathing complex.

The first room in a Roman bath was the *apodyterium*, or changing room, a space located near the entrance and usually not warmed. Benches and chairs were arranged along the walls for the use of visitors, who

View of the monumental frigidarium of the baths built by the emperor Caracalla in Rome.

The caldarium of the Stabian Baths at Pompeii. Note the basin (labrum) at the rear and the vault's grooved decoration, which served to prevent water from condensation and dripping.

could store their clothing in niches or on shelves. The Stabian Baths at Pompeii include a well-preserved example of a Roman *apodyterium*, which features a vault decorated with stuccoes of human figures and plant elements.

After passing through the *apodyterium*, a visitor to the baths generally moved next to the heated *caldarium*, which often had pools for immersion bathing (Lat. *alvei*) and fountains or basins for cold ablutions (Lat. *labra*).

The pools in this room, naturally, were relatively large and usually located at the end of the space or in deep recesses, where one could bathe by descending a series of masonry steps. After bathing in the *caldarium* a visitor went to the *tepidarium*; this room, unlike the preceding one, did not receive any direct heat. The *tepidarium* served as an intermediate space, through which the visitor reached the *frigidarium*, usually the largest covered room in the complex. Here, too, were immersion pools and basins of water, this time unheated; more sizeable complexes were also connected to a *natatio*, or large pool. In Rome, for example, each of the principal imperial bathing complexes had a large pool, most of which were decorated with a monumental nymphaeum embellished with statues and water displays.

Aside from the areas described above, baths could also be furnished with other spaces, of diverse sizes and designed for various activities. Among these the largest type was the palaestra, where visitors to the baths could stroll and perform gymnastic exercises. This space was usually located between the *apodyterium* and the *frigidarium*. Within the bathing complex there might also be a sauna (Lat. *laconicum*), the decongesting effects of which were praised by the leading Roman doctors, including Celsus and Galen. Usually circular, the sauna had a conical or domed roof with a central aperture (Lat. *lumen*) that was opened to admit light or closed with a bronze disk (Lat. *clipeus*). Long benches were arranged along the walls, but there were no tubs or basins. Another structure often included in a bathing complex was the latrine, situated at a peripheral location. A complex might also encompass exedrae, libraries, *auditoria* (public reading rooms), and halls where precious statues were displayed.

HEATING SYSTEMS IN BATHS

Until late in the republican period, rooms in baths were heated with simple braziers. Some sources credit a certain L. Sergius Orata with invention of the heating system that

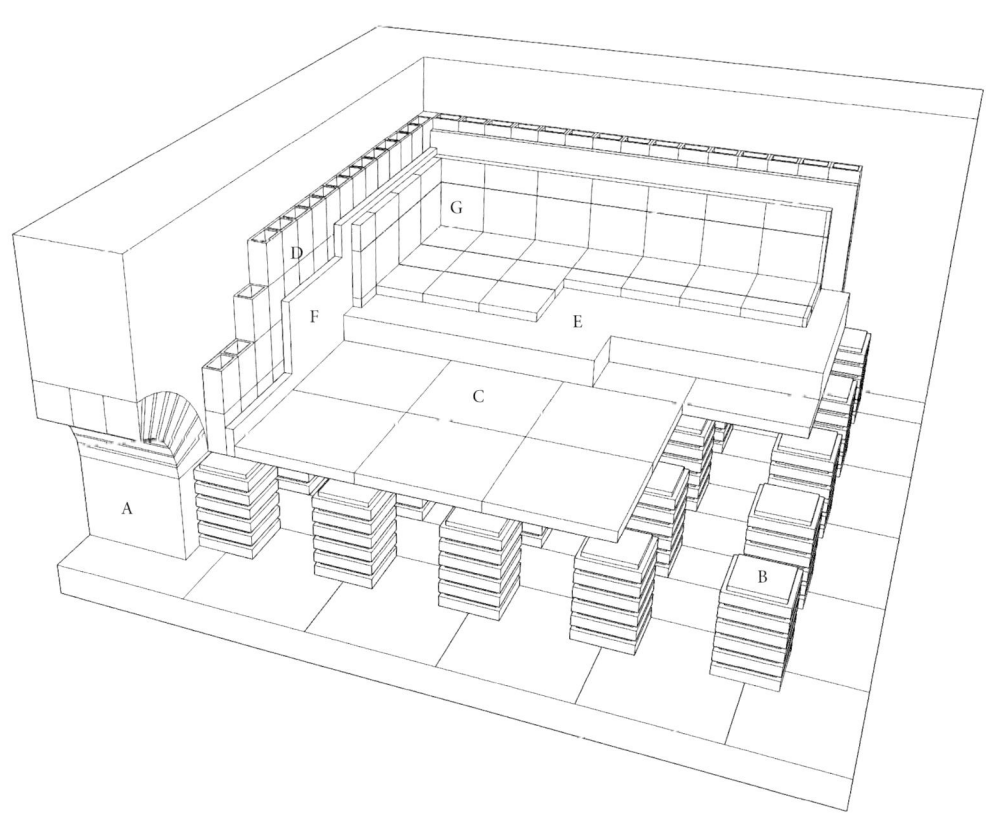

*The heating system in Roman baths: A. Passageway from the furnace (*praefurnium*); B. Pillar (*pila*) of the hypocaust; C. Series of large bricks that form part of the suspended floor (*suspensurae*); D. Hollow bricks (*tubuli*) for heating the walls; E. Floor covering in earthenware fragments; F. Wall covering in earthenware fragments and arrangement of the wall slabs; G. Marble slabs (*crustae*) that dress the floor and walls.*

distinguished all Roman baths. This well-to-do businessman, known for breeding and commercializing oysters from Lucrine Lake, was said to have been inspired by his observations of natural heating in the area of the Phlegraean Fields, which he then reproduced artificially.

The system was relatively simple and based on the passage of warm air through spaces created below floors and along walls. It came to be called a hypocaust from a technical term of Greek origin, *hypocaustum*, meaning "heating from below," an apt description of the empty space below the floors of the baths. Heat was produced by lighting a fire in a designated place, known as the *praefurnium*, located immediately next to the space to be heated. Warm air moved under the floor of the *hypocaustum*, and then passed through an arched opening that was usually flanked by two small parallel walls, upon which rested metal containers for heating water. Vitruvius (5.10.1) describes the arrangement and functions of these containers: "Three bronze tanks are to be placed above the furnace: one for the hot bath, a second for the tepid bath, a third for the cold bath. They are to be so arranged that the hot water which flows from the tepid bath into the hot bath may be replaced by a like amount of water flowing down from the cold into the tepid bath."

In these heated spaces visitors to the bath walked on a suspended floor (Lat. *suspensurae*) that was raised off the ground by means of short pillars (Lat. *pilae*) arranged to form a regular grid. In most cases these pillars were constructed of square or circular bricks, but

Remains of the hypocaust heating system with circular pillars in the imperial baths at Grumentum in Basilicata.

archaelogists have found frequent instances where the most varied materials were oddly reused to serve this purpose, from terracotta pipes to fragments of columns. Vitruvius (5.10.2) offers these specific instructions for construction of a *hypocaustum*: "The hanging floors of the hot baths are to be made as follows: first the ground is to be paved with eighteen-inch tiles sloping toward the furnace, so that when a ball is thrown in it does not rest within, but comes back to the furnace room of itself. Thus the flame will more easily spread under the floor. On this pavement, piers of eight-inch bricks are to be built at such intervals that two-foot tiles can be placed above. The piers are to be two feet high. They are to be laid in clay worked up with hair, and upon them two-foot tiles are to be placed to take the pavement." The floor itself, therefore, was made with bricks 59 centimeters on each side (Lat. *bipedales*); these were usually arranged in double layers and in such a way that four adjacent corners rested on top of a pillar. To these bricks were added one or more layers of ground pottery of various particle size, after which a dressing of marble or mosaic slabs was applied.

Of course, it was not advisable that the heat, or more precisely the warm gases, of a bathing room remain blocked under the floor. The Romans soon devised a solution that turned this problem into an advantage: within the terracotta elements they created spaces that allowed the warm gases to pass inside the walls and up to the roof. These elements were usually true chimney flues, constructed from hollow clay pieces with rectangular bases. In other cases a sort of counterwall was made using tiles with conical projections or flanges (Lat. *tegulae mammatae* or *hamatae*) that formed cavities for the hot gases. This counterwall was then dressed with attractive ground pottery for a decorative finish.

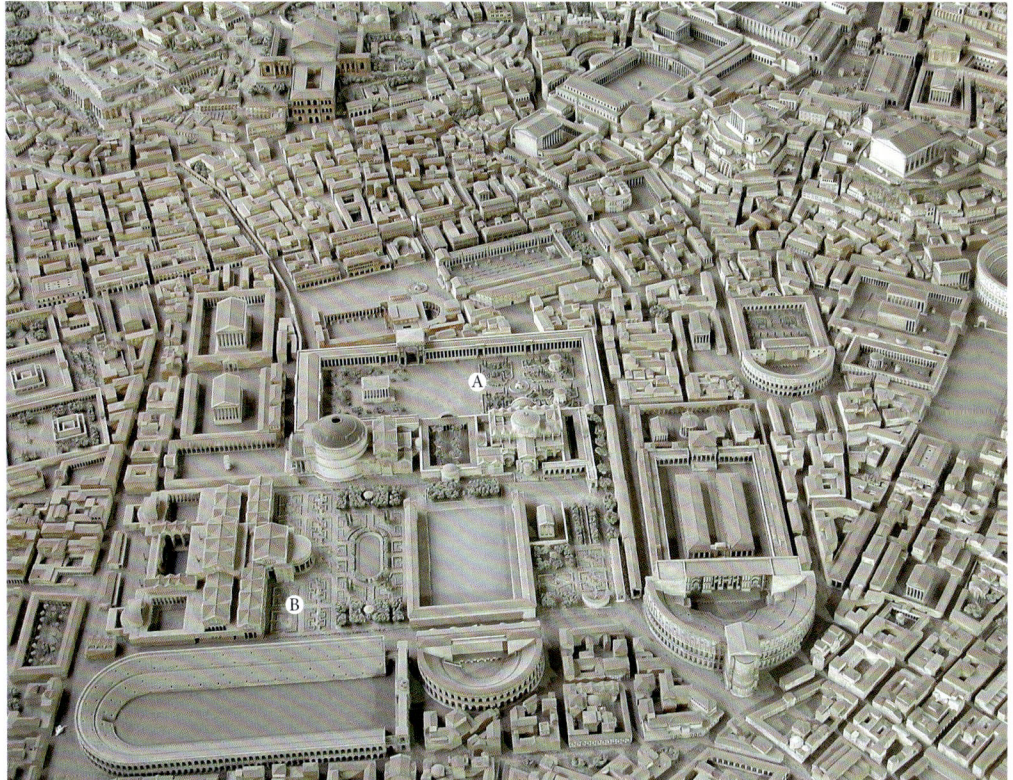

The two largest bathing complexes at the Campus Martius, as reconstructed in the model of ancient Rome at the Museo della Civiltà Romana: A. Baths of Agrippa; B. Neronianae-Alexandrinae baths.

ROME'S GREAT PUBLIC BATHS

Little remains of Rome's most ancient bathing complexes, and some of them are known only through references in literary sources. However, we are fortunate to have the remains of the bathing structures within the great circular hall at the Campus Martius. These structures gave the name to the small street that crosses them—the Arco della Ciambella ("Arch of the Ring")—and comprised the public baths built by Agrippa between 25 and 19 B.C., at the same time as building was underway on the Aqua Virgo. Subject to various restorations during the first and second centuries A.D., the complex extended approximately 100 by 120 meters and was organized with spaces arranged irregularly around a large circular hall. Although few structures have survived, it has been possible to reconstruct at least part of the outline of the baths thanks to the depiction of the monument in the *Forma Urbis Romae*, as well as on the basis of several drawings from the Renaissance. Alongside the structure was the Stagnum Agrippae, an artificial lake connected to the Tiber by the Euripus, a drainage channel that crossed the entire Campus Martius.

Farther north, still at the Campus Martius but beside the more ancient structure of the Pantheon, the emperor Nero ordered new public baths to be built around A.D. 62. Known as the Thermae Neronianae—and as the Thermae Alexandrinae following restorations carried out by Alexander Severus in A.D. 227—this complex was probably the first example of a great imperial bath complex, characterized by a symmetrical arrangement of rooms along a main axis. Some remains of the building can be found beneath the current Palazzo Madama, but its layout is known from drawings made by the Venetian architect Andrea Palladio in the sixteenth century. These were monumental baths, and included a large *frigidarium* with

Remains of one of the monumental exedrae that formed part of the baths built by Trajan in A.D. 104–9.

bathing pools that opened on a scenographic *natatio*. That this complex was truly enormous is indicated by the size of its monolithic granite-column shafts, which reached a height of 11.8 meters. Three of these shafts were reused in the pronaos of the Pantheon.

After Nero's death the emperor Titus inaugurated great public baths in A.D. 80 on the site of the Domus Aurea; it is possible that their construction reused, in part, the very structures of Nero's palace. This monument too is known thanks to a drawing by Palladio, who depicts it with a symmetrical layout organized along a central axis. From the area of the Colosseum a monumental stairway reached a vast square, at the center of which stood the heated rooms of the baths. These spaces were connected by way of the *tepidarium* to a vast *frigidarium*, covered with cross vaults and flanked by two palaestrae.

A few years later, between A.D. 104 and 109, Trajan assigned his architect Apollodorus of Damascus the task of building a new bathing complex immediately northeast of the Thermae Titianae, which also were located in the area of the former Domus Aurea. Fed by the Aqua Traiana and the great water cistern of the Sette Sale, the Thermae Traianae offer the first example of monumental baths in which the central structure of the bathing rooms is elevated inside a large enclosure with a large semicircular exedra. Moreover, Apollodorus chose to orient the building in such a way that the caldaria profited from the maximum amount of sunlight between noon and sunset, which thus contributed to the natural heating of the rooms. As noted earlier, this design had been suggested in the first century B.C. by Vitruvius (5.10.1); it was also adopted in the great imperial complexes later built by Caracalla, Diocletian, and Decius.

The central body of Trajan's baths measured 190 by 212 meters, but with the addition of the enclosure the complex extended to 330 by 315 meters. The monument has been the object of recent research into the technical aspects of the structures, and archaeologists have determined that in its northern area a large *natatio* was flanked by groups of rooms, each of which was joined to a large circular hall.

The Baths of Trajan represent an architectural complex of monumental dimensions, but they were soon outdone by the baths that Caracalla constructed from A.D. 212 through 217 in the southern area of the city, between the Via Ardeatina, the Via Appia, and the Via Latina. To deliver water to these baths the emperor ordered construction of a new stretch of the Aqua Marcia, later called the

View of the remains of the Baths of Caracalla.

Aqua Antoniniana. Still well preserved, the monument reimagined the enclosed model of the Thermae Traianae, but at an even more colossal size (337 by 328 meters). The central structure, which alone measured 220 by 114 meters, once again was given a symmetrical contour, with the axial arrangement of a vast *natatio*, *frigidarium* (or *basilica thermarum*), *tepidarium*, and grand circular *caldarium*. To either side were the palaestrae, around which were organized groups of rooms, some heated, of varying function. The organization of the enclosure was equally monumental, with gardens, porticoes, exedrae, and libraries. Hidden by the stairs of a sort of stadium were the enormous western cisterns, whose capacity reached 89,000 cubic meters of water. The overall structure was served by an underground network of streets, to which all the service rooms were connected. A gigantic network of canals, conduits, and sewers spread under these streets and flowed into a large sewer located on the southwestern side of the complex.

Under the emperor Decius (A.D. 249–51), additional baths were constructed on the Aventine. It was Diocletian, however, who ordered the largest bathing complex in Rome, the Thermae Diocletiani. Built in the area of the Castra Praetoria to serve the residential quarters that had developed just beyond the Quirinal, Esquiline, and Viminal hills, the baths were completed over a course of only eight years, between A.D. 298 and 306. They extended 380 by 370 meters, and adopted a plan based on an

The monumental structures of the Baths of Diocletian are still well preserved due to their reuse in the complex of the church of Santa Maria degli Angeli.

enclosure with exedra and a central body of 250 by 180 meters; the central space is today occupied in part by the church of Santa Maria degli Angeli. Fed by a branch of the Aqua Marcia and served by a large trapezoidal cistern (the Botte di Termini), the bathing nucleus of the Thermae Diocletiani repeated the plan that had been adopted by emperors since Nero: an axial succession of main spaces (the *natatio*, *frigidarium* or *basilica thermarum*, *tepidarium*, and *caldarium*) with palaestrae, atriums, and dressing rooms in a lateral symmetrical arrangement. Along the enclosure ran porticoes, exedrae with fountains, meeting rooms, and public auditoriums, while on the external perimeter of the southern side stood two large circular halls, one of which is today the church of San Bernardo alle Terme.

ROADS, BRIDGES, AND TUNNELS

GREEK ROADS

By no means can the Greeks be considered great builders of roads. The sea was their vocation; they traveled by ship, not only because of the geography of the Hellenic world—composed of an infinity of large and small islands—but also due to the mountainous landscape of their territory. Nevertheless, archaeologists have identified the remains of roads built as early as the second millennium B.C. on both Crete (at Lasithi) and the Peloponnesus (at Arcadia, Argolid, and Messenia), as well as in continental Greece (at Attica, Boeotia, and Phocis). In some cases these roadways were wide and well built, and occasionally they were terraced and supplied with a drainage system.

It appears that during the historical period the Greeks did not continue this type of construction, at least not for communication routes outside cities. Pausanias (*Desc. of Greece* 2.11.3) writes that, as late as the second century A.D., many roads—such as the one to Titane in the Corinthia—were "too narrow to be used by carriages drawn by a yoke." Referring to the route between Cleonae and Argos (2.15.2), he describes two roads, both barely

View of the plain of Corinth, reached by the Scironian Road from Megara.

Route of the Diolkos between the Gulf of Corinth and the Saronic Gulf.

adequate: "One is direct and only for active men, the other goes along the pass called Tretus ['Pierced'], is narrow like the other, being surrounded by mountains, but is nevertheless more suitable for carriages." However, the situation was not always so unfavorable for travel. Pausanias (1.44.6) also mentions the quality of ancient roads that joined Megara to Corinth, which had been reworked: "The road called Scironian to this day and named after Sciron, was made by him when he was war minister of the Megarians, and originally they say was constructed for the use of active men. But the emperor Hadrian broadened it, and made it suitable even for chariots to pass each other in opposite directions."

In the Greek world roads between cities, and those between individual cities and the countryside (Gr. *chora*), were usually made simply by leveling the area of the thoroughfare to make it passable by carts. Often, however, sacred routes that led to great sanctuaries located outside a city were paved with large blocks of stone, as was the case with the sacred complexes at Delphi, Eleusis, and Olympia.

Roads serving quarries represent a special type of construction, since they were designed for the transport of blocks of stone, as on Mount Pentelicon as well as on the island of Kos. These roads were paved with shards and blocks of quarry material to reduce friction on the sledges with which stone was transported, thus facilitating their passage.

THE PORTAGE ROAD AT CORINTH

Before the creation of the Corinth Canal at the end of the nineteenth century, the Peloponnesian peninsula was united to mainland Greece by an isthmus just over 6 kilometers long, located between the Gulf of Corinth and the Saronic Gulf. Ships transporting goods between the Ionian and the Aegean seas were obliged to sail along the Peloponnesian coast. Pliny the Elder (*Natural History* 4.4) describes the austere terrain of this region: "The Peloponnesus . . . is a peninsula . . . between two seas, the Aegean and the Ionian. . . . The narrow neck of land from which it projects is called the Isthmus. At this place the two seas that have been mentioned encroach on opposite sides from the north and east and swallow up all the breadth of the peninsula at this point, until in consequence of the inroad of such large bodies of water in opposite directions the coasts on either side have been eaten away so as to leave a space between them of

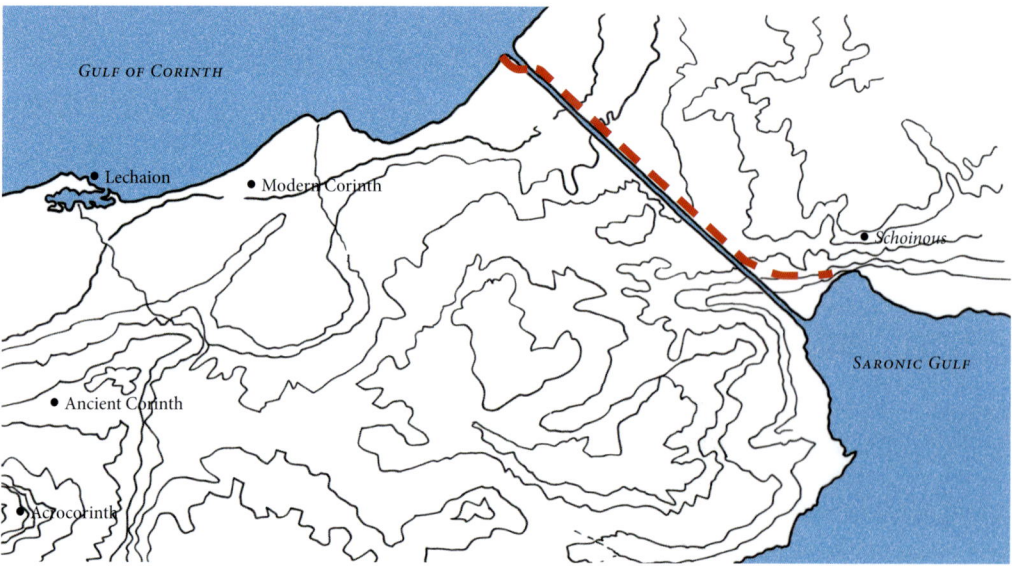

only five miles." He further describes the unsuccessful attempts to make a passageway for ships from one side of the peninsula to the other: "The circuit of the Morea [Peloponnesus] is a long and dangerous voyage for vessels . . . and consequently successive attempts were made by King Demetrius, Caesar the dictator, and the emperors Caligula and Nero to dig a ship canal through the narrow part—an undertaking which the end that befell them all proves to have been an act of sacrilege."

The concept of digging a canal across the isthmus is ancient, dating back even before the rule of Demetrius I of Macedon, known as Poliorcetes ("the Besieger," 336–283 B.C.) and mentioned in the passage from Pliny cited above. In fact, the tyrant Periander had planned such an undertaking at the end of the seventh and beginning of the sixth centuries B.C., finally abandoning the project for fear it would do more financial harm than good to Corinth. It was Nero who began the excavation in the middle of the first century A.D., using six thousand slaves from Galilee. The work was planned for a width of about 50 meters and a depth of 10 meters, but remained unfinished after covering less than 2 kilometers. Pausanias (*Descr. of Greece* 2.1.5), for whom the signs of this enterprise were still fresh, writes, "Where they began to dig is still to be seen, but they did not advance at all into the rock."

The Greeks had meanwhile circumvented this problem with a singular system: they opened a road between the two seacoasts and used it for portaging the cargoes of ships, or even the ships themselves, loading them onto carts. Greek sources refer to this route as the Diolkos, a term that according to Strabo (*Geography* 8.2.1) indicates the place where "ships are hauled overland from one sea to the other." This portage route was still in operation during the fifth century B.C. and the years of the Peloponnesian War. Thucydides (*History of the Peloponnesian War* 3.15.1) reported that the Spartans and their allies "proceeded to construct on the Isthmus hauling machines with which to transfer the ships from Corinth to the sea on the Athenian side." He later adds (8.8.3–4) that the Spartans first transported half of the ships across the Isthmus: "These were to set sail immediately, in order that the attention of the Athenians might not be directed toward the ships that were setting out more than toward those that were afterward being carried across the Isthmus. . . . They at once conveyed twenty-one ships across."

Archaeological excavations have uncovered several short stretches of the road built by the Spartans, paved in local stone with a width between 3.6 and 5 meters. Still preserved in the roadway are a pair of parallel grooves, about 1.5 meters apart, that served as guides for the wheels of the carts. The route of the road began at the Gulf of Corinth just west of the site of the future canal and proceeded in a straight line along the eastern side, curving to the east to reach the Saronic bank at Schoinous, near the modern village of Kalamata. This route made it possible to cross the isthmus at the

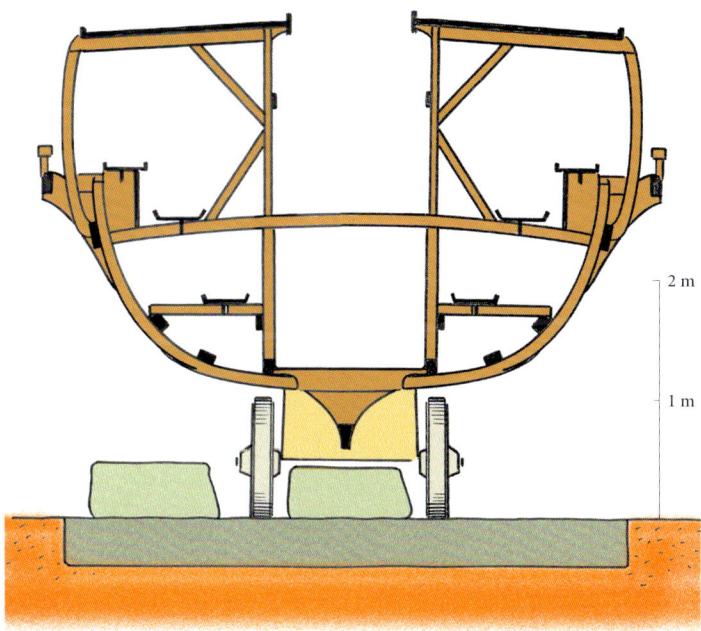

Hypothetical reconstruction of the method for transporting ships along the Diolkos, proposed by D. Werner.

The great canal in stone blocks built at the end of the fourth century B.C. on the edge of the plateia *that separated the agora of Metapontum from the urban sanctuary.*

lowest altimetric point, located at an altitude of only 75 to 80 meters above sea level.

These transportation operations cannot have been easy, given that triremes were about 35 meters long and weighed 20 to 30 tons. Scholars have proposed that to move such a ship it was probably set on a large wooden platform with wheels, about 3.5 meters wide and 20 meters long. The platform was then dragged by approximately 120 people. It would have been impossible to manage an equivalent animal force, calculable at 80 to 100 pairs of animals.

ROADS IN GREEK CITIES OF THE WEST: METAPONTUM AND THURII

The shape of a Greek city often depended on its urban history. Many of the most ancient of the *poleis*—Eretria on Eubea, or Athens itself—began as a collection of small separate villages (Gr. *kata komas*). The social and political aggregation of these communities usually resulted in the formation of an irregular road network based on spontaneous routes that were more or less regularized. In contrast, cities created through a deliberate act of foundation (or refoundation) had more-methodical road networks, the fruit of precise plans. In general, public and religious areas, as well as those destined for the homes of citizens, were defined by a layout of roads that reflected a precise hierarchy: There were roads (Gr. *plateiai*) with a width of as much as 30 meters, and narrow roads (Gr. *stenopoi*). *Plateiai* and *stenopoi* crossed one another perpendicularly,

creating a grid that determined the shape and size of individual city blocks. The examples below will help to clarify the relationships between these two types of roads.

Metapontum was founded by the Achaeans in the final decades of the seventh century B.C., along the Ionic coast in what is now Basilicata, between the Basento and Bradano rivers. The urban space reached a size of about 150 hectares, and was defined by the routes of the two riverbeds as well as by the coastline. During the colony's first stage of life, the area inside the walls appears to have been divided into public areas (including the agora and sanctuaries) and residential areas by a single urban road, about 22 meters wide, which ran at a right angle to the coast. Other roads were laid out later to form a network of perpendicular routes based on at least three large *plateiai* running north to south and two *plateiai* running east to west. Parallel to the east–west roads were *stenopoi*, about 5 to 6 meters wide, that formed long blocks with a constant width of 35 meters.

Connected to this system of roads was a method for the runoff of water, based initially on canals dug to the sides of the *stenopoi*. Only later, around the middle of the fourth century B.C., was a series of masonry canals built. By way of various discharges these conduits received water from individual blocks and conveyed it to large limestone structures for collection that were built along the *plateiai*.

Farther south, also along the coast of Magna Graecia, was the Achaean city of Sybaris, whose settlers had founded Metapontum. The *polis*, after reaching the apex of its power in the sixth century B.C., was destroyed by the Crotoniates in 510 B.C., but the area was not abandoned. On the same site in 443 B.C., the Athenians of Pericles founded Thurii, entrusting the design of its layout to the great Hippodamus of Miletus. This Greek architect and urban planner, already responsible for the planning of Piraeus, organized Thurii along a system of orthogonal roads composed of four *plateiai* arranged laterally and crossed at right angles by another three. The urban surface was thus divided into at least six large areas, each further subdivided into smaller blocks (about 35 by 72 meters) by a grid of orthogonal *stenopoi*. A sewer, 1.8 meters wide, ran at the center of each *stenopoi*, to form lots of about 35 by 35 meters, each of which was perhaps designed to receive two homes.

Top. View of the excavations at Sybaris, with remains of monuments from the Roman period of the site.

Bottom. The great north–south plateia *of Thurii, one result of the city planning of Hippodamus of Miletus.*

Right. Detail of the model of the reconstruction of Rome in the fifth century A.D. displayed in the Museo della Civiltà Romana in Rome.

Opposite. Stretch of the Via Sacra between the Arch of Titus and the Forum Romanum.

Archaeological excavations have brought to light some portions of this road system, which remained active—although various modifications were made to the width of the streets—even after the Romans founded the Latin colony of Copia there in 194 B.C. These roads had been constructed with particular care, paved with large slabs of stone that sloped to the sides, allowing rainwater to run off into canals built along the sides of the roads or set beneath them.

It is not only because of the connection to Hippodamus that the roads of Thurii have assumed a certain importance. Diodorus Siculus (*Historical Library* 12.10.7) notes the name given to each of the city's great *plateiai*: the four running lengthwise were called Heracleia, Aphrodisia, Olympias, and Dionysius; the other three were Heroa, Thuria, and Thurina. Some of these roads probably passed a sanctuary and were assigned the name of the divinity to which the sacred area was dedicated. This method of identifying streets appeared in other cities of the Greek world, such as on Thasos, where there was a "road of the sanctuary of Herakles" and a "road of the sanctuary of the Charites." But it was not always thus: the main street of ancient Cyrene in Libya was called Skyrota because its roadbed was made of gravel.

ROMAN ROADS

From its very beginnings Rome developed in a strongly strategic location. The first lasting settlements, which date to long before the mythical foundation of the city by Romulus in 754/53 B.C., arose near the site where it was easiest to ford the Tiber—near Tiber Island, almost at the foot of the Capitoline. These settlements soon led to the formation of an area used for markets and exchanges with nearby communities,

a site that survived throughout the city's history in the place names of the Forum Holitorium ("Vegetable Market") and Forum Boarium ("Cattle Market"). This important area made its way into myth as the site of the battle between Hercules and Cacus, after which the hero crossed the Tiber, driving ahead of him the sacred cattle of the giant Geryon.

The Romans always took special care of their roads. As early as the ancient laws of the Twelve Tables, the first written code of Roman law, there were rules concerning the roads of the city (Lat. *urbs*), with indications of their dimensions and restrictions on their usage. Thus we know that in the fifth century B.C. laws were passed regulating the passage of cattle over streets paved in stone. Several centuries later the Greek geographer Strabo (5.3.8) expressed admiration for the Romans, who paid greater attention to the repair and maintenance of their roads than did the Greeks: "If the Greeks had the repute of aiming most happily in the founding of cities, in that they aimed at beauty, strength of position, harbors, and productive soil, the Romans had the best foresight in those matters which the Greeks made but little account of, such as the construction of roads and aqueducts, and of sewers that could wash out the filth of the city into the Tiber. Moreover, they have so constructed also the roads which run throughout the country, by adding both cuts through hills and embankments across valleys, that their wagons can carry boat loads; and the sewers, vaulted with close-fitting stones, have in some places left room enough even for wagons loaded with hay to pass through them."

We know at least forty-one street names from ancient Rome. Some of them, such as the Clivus Capitolinus and the Clivus Argentarius, contained a reference to the slope (Lat. *clivus*) along which they ran; others, such as the Vicus Iugarius and the Vicus Longus, indicated the road of a specific residential quarter. Added to these were the *viae*, such as the Via Lata ("Broad"), the urban portion of the great road to Rimini, and the Via Nova, located on the north side of the Palatine. The most ancient and important of all the streets was the Via Sacra, along which moved formal religious and other processions that terminated on the Capitoline near the Temple of Jupiter Capitolinus. The best-preserved section of this road is that between the Arch of Titus and the Regia, belonging to the Augustan phase of construction. Its large paving stones were discovered when nineteenth-century excavators destroyed the higher level of pavement—which had been laid down following the fire of A.D. 64—mistaking it for medieval work. Following this excavation, sections of the monument near the Via Sacra that had been constructed on the level of the later pavement—including the Arch of Titus, the Basilica of Maxentius, and the Temple of Romulus—were left with their foundations above ground.

ROADS OF ROMAN ITALY AND THE PROVINCES

Livy (*History of Rome* 5.54.4–5) synthesizes the strategic position at the base of Rome's fortune: "Not without cause did gods and men select this place for establishing our City—with its healthful hills; its convenient river, by which crops may be floated down from the midland region and foreign commodities brought up; its sea near enough for use, yet not exposing us, by too great propinquity, to peril from foreign fleets; a situation in the heart of Italy—a spot, in short, of a nature uniquely adapted for the expansion of a city." The city sprang up where the navigable course of the Tiber intersected with the most ancient road connecting the Tyrrhenian coast to the inland areas of Etruria, Latium, and Campania. The Via Campana connected the salt flats (Lat. *campus salinarum*) at the mouth of the Tiber to Rome, and then continued by the Via Salaria, which from the urbs crossed the Apennines to the Adriatic, acting as the vector for commerce in salt, a necessary product for the production and preservation of food.

Remains of two columns that indicated the terminus of the Via Appia at Brindisi.

As Strabo (*Geography* 5.3.9) noted, roads also performed a fundamental role in Rome's military expansion and in the later organization of Rome's territory, beginning in the region of Latium itself: "As for the rest of the cities of Latium, their positions may be defined, some by a different set of distinctive marks, and others by the best-known roads that have been constructed through Latium; for they are situated either on those roads, or near them, or between them." A true network of roads came into existence around Rome, some of which still survive with various modifications.

In general, each road was named for the city toward which it was directed. Thus there were the Via Nomentana, which went to Nomentum (near modern Mentana), the Via Tiburtina to Tibur (Tivoli), the Via Collatina to Collatia (perhaps identifiable with the area of Lunghezza), the Via Gabina-Praenestina, which originally ran to Gabii but was then extended to reach Praeneste (Palestrina), the Via Labicana to Labicum (located in the Valle del Sacco), the Via Ardeatina to Ardea, and the Via Laurentina to Laurentum (near Castel Porziano). In addition, the Via Ostiense led to the mouth (Lat. *ostium*) of the Tiber, where the fortifications (Lat. *castrum*) of Ostia arose early in the fourth century B.C.

With the advance of Rome's conquests and the foundation of colonies throughout the Italic peninsula, the Romans began to lay out large

Stretch of the Via Appia near the villa of the emperor Maxentius.

roads that would connect the new territories to the capital. Among these the Via Appia (Appian Way), begun in 312 B.C. by the censor Appius Claudius Caecus, was the first to be given the name of its builder. The road joined Rome to Capua (modern Santa Maria Capua Vetere) but was soon extended, first to Benevento, then to Taranto, and subsequently to Brindisi, the bridgehead for connections to Greece and the Orient. Capua was the starting point for a road to Rhegium (Reggio Calabria), perhaps built by Titus Annius Rufo in the second half of the second century B.C., while in Sicily two large coastal roads began in Messina, one heading north toward Lilybaeum (Marsala), the other heading south toward Syracuse.

As for northern Italy, a passage in Cicero (*Philippics* 12.9) relates that three main consular roads—so called because they bear the names of consuls—went to Rome: the Flaminian (modern Via Flaminia), the Aurelian (Via Aurelia), and the Cassian (Via Cassia). The first had been opened by Gaius Flamminius around 220 B.C. to reach the Adriatic coast at Rimini, passing through Narni, Foligno, and Fano; the second was made by members of the Aemilia family to establish a connection to the coastal cities of Etruria; the third was built by Cassius Longinus to reach Florentia (modern Florence) and from there perhaps Luna (Luni). To these three roads can be added, among others, the Via Aemilia, between Piacenza and Rimini; the Via Postumia, between Genoa and Aquileia; and the Via Annia, also directed to Aquileia.

The Romans are also known to have built important communication roads to connect provinces. Among the oldest and most important are the Via Domitia, which was built by the proconsul Domitius Ahenobarbus around 121 B.C. to link Italy and Spain, and the Via Egnatia, which crossed Illyria and Macedonia to reach Salonika.

All these roads were built at the expense of the state, which was also responsible for their maintenance. There are numerous testaments to the restoration works promoted by emperors, most importantly those of Augustus, but also those of Claudius, Vespasian, and Trajan. Upkeep of the road network was entrusted to officials explicitly in charge of road conditions (Lat. *curatores viarum*), and use of the roads was regulated by specific laws. For example, the *Digest of Justinian*, part of the "Body of Civil Law" issued around 529–34 A.D., includes prohibitions against building, digging, and dumping earth on roadbeds; it was also forbidden to deposit garbage there or to occupy the space. Exceptions to these rules included certain classes of artisans, such as those who dyed cloth (Lat. *fullones*) and the makers of carts (Lat. *fabri*): the first were permitted to hang up their fabrics to dry, while the second had the right to display wheels for sale, but only if the sellers took care not to obstruct the passage of vehicles.

From a legal and administrative point of view, not all roads were equal. Principal roads (Lat. *viae publicae*) were distinct from minor roads, such as the crossroads (Lat. *viae vicinales*) that not only connected villages to principal roads but also connected villages to each other. In addition, there were private roads (Lat. *viae privatae*), built by individual landowners on their property, and military roads (Lat. *militares*), constructed by armies exclusively for strategic uses.

ROMAN ROAD-BUILDING TECHNIQUES

The immense network of urban and intercity roads constructed by the Romans throughout the empire constitutes the outcome of a long, evolving process, one based on experiments and the application of advanced engineering. As early as the fourth century B.C., the techniques of laying roads were highly advanced, as indicated by a stretch of the Via Lavinate that has been discovered in the area of Laurentino, on the periphery of Rome. The road, 2 to 3.5 meters wide, was created by digging into the tufaceous rock and creating a roadbed of stone. Built to each side was a curb made of large fragments of tufa, while other blocks facilitated the draining of rainwater and perhaps also served as sidewalks.

A passage composed in the first century A.D. by the Roman poet Statius (*Silvae* 4.3.40–49) is the only ancient source that provides any details of Roman road-building techniques. It refers to the Via Domitiana, inaugurated by Domitian in A.D. 95 to connect the Via Appia to Naples by way of Cuma and Pozzuoli: "The first task here was to start on furrows and cut out borders and hollow out the earth far down with a deep excavation. Next, to fill the trenches they dug with other material and prepare a basin for the raised spine, so that foundations do not wobble nor a niggardly bottom offer a treacherous bed for the packed stones. After that, to knit the road with blocks close set on either side and with frequent wedges. Oh, how many hands work in unison!"

More generally, the construction of a road called for the preparation of flat ground and the creation of a first layer of cobblestones, which served to compact the earth and drain away water. Over this layer was placed first a thick layer of sand and gravel and then a covering mantle of stones, slabs, or large paving stones of hard rock. A roadway was often accompanied along the edge by a sidewalk or a canal for the collection of rainwater. The construction of some roads appears to have corresponded to the terminology given by Vitruvius (7.1.3) with regard to the composition of architectural floors: The first layer would thus correspond to

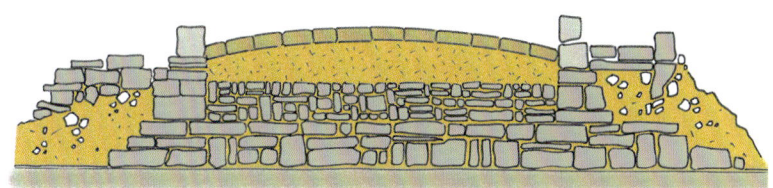

Top. Cross-section reconstruction of the preparatory layers of a Roman road.

Bottom. Reconstruction of the building system with wooden subfoundations for the roadbed of the Via Mansuerisca, along the stretch that crossed the swampy area of Hautes Fagnes in Belgium.

the *statumen* (the foundation), the second to the *rudus* (a layer of coarse concrete), and the third to the *nucleus* (a layer of fine concrete), which was topped by the *summum dorsum* (a final surface of large stone slabs). When the terrain was soft or swampy, the Romans reinforced it with a sort of wooden framework composed of lengthwise and crosswise planks that were anchored to other wooden elements fixed in the ground.

The construction of many of the consular roads reflects the techniques described above, often with the addition of further refinements. Some stretches of the Via Appia, for example, were made with a layer of sand and gravel that measured about a meter in depth, on which rested a second, thinner layer, either composed of gravel mixed with bits of limestone or composed of pieces of limestone alone. The Via Aurelia was given a compact roadbed in tufa blocks that were covered with gravel and stones to a height of 40 centimeters.

The materials used in the construction of roads were, as always in the ancient world, those available locally: limestone, tufa, and volcanic stone in Latium and central Italy; limestone and sandstone in the Adriatic regions and Sicily; gravel and pebbles in the north. The Romans might have been expected to apply lime mortar as a binder to these materials, but instead they preferred to use clay and well-compacted earth.

Roads were given different names according to the type of roadbed. Some, such as the *viae terrenae*, were left with a simple surface of nothing more than beaten and leveled earth; others, such as the *viae glarea stratae*, had a mantle composed of a layer of gravel and pebbles; still others were carefully paved in large blocks of hard stone (Lat. *viae lapidibus stratae*) or in smaller slabs arranged in a more regular way (Lat. *viae silicae stratae*).

It was only in the second century B.C. that roads between cities were paved. A passage in Livy (*History of Rome*) reports that in 174 B.C. paving was confined to urban roads (Lat. *intra muros*), while the long stretches outside cities were simply covered with beaten earth or stones. This system was later codified in the *Digest of Justinian*, but archaeology has revealed that roads were frequently paved even in areas outside cities. For example, the Via Appia was paved up to Bovillae, located 11 Roman miles south of Rome, and then again in the stretch between Anxur Tarracina (Terracina) and Formia.

Many roads were heavily trafficked by carts laden with foodstuffs and other goods necessary to daily life. As a result of this constant wear from wheels, the passage of carts has left characteristic ruts in road surfaces. In other cases, such as routes over mountains or across particularly tortuous stretches of ground, furrows in the surface were purposely created to serve as guides, functioning like train tracks, to prevent carts from slipping off the road.

Several inscriptions recall the costs of building roads. An epigraph relating to restoration work carried out in 82 B.C. on the Via Caecilia reports an expense of 150,000 sesterces for the repair of a 20-mile stretch in the Apennine area. This would seem a costly operation, as during the same period daily wages for a worker in Rome were about 3 sesterces.

BRIDGES: THE PERSIAN EXPERIENCE AND THE GREEK WORLD

It was with the construction of bridges (Gr. *gephyra*, Lat. *pons*) that ancient architects reached the highest levels of engineering. A bridge appears to be a relatively simple structure: a passage suspended on two or more supports. Yet the creation of a bridge represents the outcome of careful planning based on the characteristics of the terrain in which the foundation is to be laid, as well as knowledge of the forces exercised by the mass of water delivered over the course of a year by the river crossed.

The need to build bridges and viaducts appeared very early in the ancient world. In the second millennium B.C., to the southwest of the palace of Knossos, Minoan architects

Left. Hypothetical reconstruction of the two bridge boats made by the Persian emperor Xerxes in 481 B.C. to transport his army across the Hellespont.

Bottom. Diagram of the three-branched bridge built at the confluence of the Mavrozoumenos and Vivari rivers near Meligalas in Messenia.

erected a bridge with what are known as false, or corbel, arches over the Vlychia stream; and a little later, in the direction of the coast, they built a viaduct, also with several false arches, that was more than 22 meters long. The Mycenaeans used much the same strategy during the second half of the second millenium, as demonstrated by several bridges with false arches in the Peloponnesus, including at Argolis.

Sources from the historical period mention the existence of numerous bridges, beginning with that built at Babylonia by Queen Nitocris in the sixth century B.C., described by Herodotus (*Histories* 1.186):

> Her city was divided into two portions by the river which flowed through the center. Whenever in the days of the former rulers one would pass over from one part to the other, he must cross in a boat; and this, as I suppose, was troublesome. But the queen provided also for this. When the digging of the basin of the lake was done . . . she had very long blocks of stone hewn; and when these were ready and the place was dug, she turned the course of the river wholly into it, and while it was filling, the former channel now being dry, she bricked with baked bricks, like those of the wall, the borders of the river in the city and the descents from the gates leading down to the river; also about the middle of the city she built a bridge with the stones which had been dug up, binding them together with iron and lead. She laid across it square-hewn logs

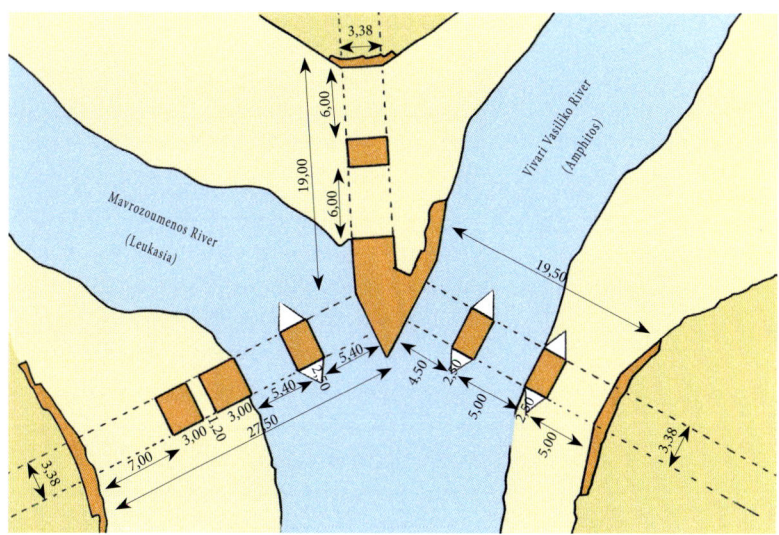

The bridge built in the base of the propylaeum erected in the third century B.C. at the entrance to the Sanctuary of the Great Gods at Samothrace.

each morning, whereon the Babylonians crossed; but these logs were taken away for the night, lest folk should be ever crossing over and stealing from each other.

One of the most ancient and distinctive types of bridges was suspended on floating ships, a method used by the Persian king Darius (522–485 B.C.) to move his army over the Bosporus. Alexander the Great (356–323 B.C.) later employed the same type of bridge over the Pyramus River in Cilicia, and over the Euphrates at Thapsacus, where he completed several bridges of boats that had been begun by the Persians. To expunge the fortress of Petra, protected on all sides by ravines, he built a large wooden viaduct by fixing poles, made from fir trees, in the ground at regular intervals and joining them with a network of rushes.

As these structures were made of perishable materials, they are known today only through ancient literature, but other bridges from the Greek period, built in stone, remain at least partially preserved. Among the oldest is the four-span bridge erected in the fifth century B.C. immediately outside the walls of Eretria in Eubea. Made partially of wood, this structure was built to replace an earlier bridge, destroyed by the Persians in 490 B.C. Around the middle of the fifth century, however, a bridge constructed entirely of stone was built at the entrance to the sanctuary of Artemis at Brauron, near the eastern coast of Attica, making it possible for the sacred road from Athens to cross over a small stream of water. The four bays of this bridge were supported on piers of squared limestone blocks. Its road surface, made with large superimposed slabs of the same material, still bears ruts from the passage of carts.

Far more complex in its structure was the three-arm bridge built at Messenia at the

Detail of one of the bridges with a false (corbelled) arch discovered at Selinunte in Sicily.

confluence of the Mavrozoumenos (ancient Leukasia) and Vivari (Amphitos) rivers. Erected over the course of the fourth century B.C., its Y shape resulted from the need to build an arm of the bridge over an alluvial cone that had formed upriver of the confluence. The hinge of the entire structure was a great central pylon, made in regular *opus quadratum* with blocks of local limestone. The system by which the bays were covered, whether a wooden structure or arches made of wedge-shaped stones, remains a mystery. The use of such stones can be seen in a small bridge built during the same century at Rhodes, near the bay of Akandia, to cross an artificial canal 2.8 meters wide. Similar stones were also used for the bridge made in the base of the propylaeum, erected between 285 and 280 B.C. by the king Ptolemy II Philadelphus of Egypt, in the Sanctuary of the Great Gods at Samothrace.

Among the bridges built by the Greeks in the West, those made in Sicily particularly merit discussion, both for their state of preservation and for their adoption of several specific construction techniques. No longer visible, unfortunately, is the structure erected at Agrigento just north of the Olympieion, between the end of the fifth century B.C. and the beginning of the sixth. This bridge employed a false arch with a nearly triangular shape, similar to that used for the bridges discovered at Selinunte along the ditch that separated its acropolis from the northern section of the city. Built in the fourth century B.C., the two bridges at Selinunte performed a civic function—connecting the two sections of the town—but also served a military purpose since they were easy to disassemble. The same defensive need was resolved at Syracuse with construction of a true drawbridge, erected in the final decades of the third century at the entrance to Castello Eurialo.

Top. Detail of the Ponte Rotto ("Broken Bridge"), heir to the ancient Pons Aemilius.

Opposite, top. Ponte Fabricio (ancient Pons Fabricius), built to connect the Tiber Island to the left bank of the Tiber River.

Opposite, bottom. Inscription on one of the arches of the Pons Fabricius, bearing the name of Lucius Fabricius, who built the bridge in 62 B.C.

BRIDGES OF ROME

As we have seen, the history of Rome is intimately tied to the Tiber, for it was precisely at the site of an easy ford, facilitated by Tiber Island, that the city arose. The city and the river form an indissoluble pair, as confirmed by construction of the first bridge known from ancient sources, the Pons Sublicius, located near the Forum Boarium opposite Hercules' altar, the Ara Maxima, and made by Ancus Martius (641–616 B.C.), Rome's fourth king. Inserted between two other works ordered by Ancus Martius, the bridge was made of wood, possibly oak. This material later came to be of strategic importance, when in 507 B.C. the Romans found it necessary, and easy, to destroy the bridge, while on the other side of the river Horatius Cocles held at bay the Etruscans of Lars Porsenna, determined to attack the city.

The Pons Sublicius was also sacred, its care and maintenance entrusted directly to the priests (Lat. *pontifices*). An oracle had prohibited the use of any reinforcements, including iron or bronze nails, in the bridge's construction, and thus it was built exclusively of wooden beams and pilings (Lat. *sublicius*). The bridge was also the site of an important religious ritual, that of the Argei, which took place on a day corresponding to the modern May 14 when twenty-four effigies, made of rushes and depicting men with their hands tied, were cast into the river. According to some sources, this ceremony expressed gratitude to the god Tiber for having accepted the "yoke" of the bridge.

Rome waited centuries for a bridge to be erected from permanent materials. It was only in 179 B.C. that, according to Livy (*History of Rome* 40.51.4), construction began on the first bridge with stone pylons, the Pons Aemilius, located immediately downstream from the Pons Sublicius. However, its stone arches were made a few decades later, as part of an undertaking by Publius Scipio Africanus and Lucius Mummius, who replaced the original wooden structure, perhaps in 142 B.C. Known today as Ponte Rotto (following reconstruction in the sixteenth century), it was built through a skillful combination of squared blocks of different stones: Grotta Oscura tufa, Aniene tufa, travertine, and gabina.

Following construction of the Milvian Bridge in 109 B.C., Lucius Fabricius in 62 B.C. made the first stone bridge, the Pons Fabricius, between the eastern bank of the Tiber and Tiber Island, to facilitate passage of the Via Flaminia over the river north of the city. Still well preserved, the structure is 80 meters long and 6 meters wide, and was built with travertine as well as blocks of pietra sperone and Aniene tufa. It has two large round arches

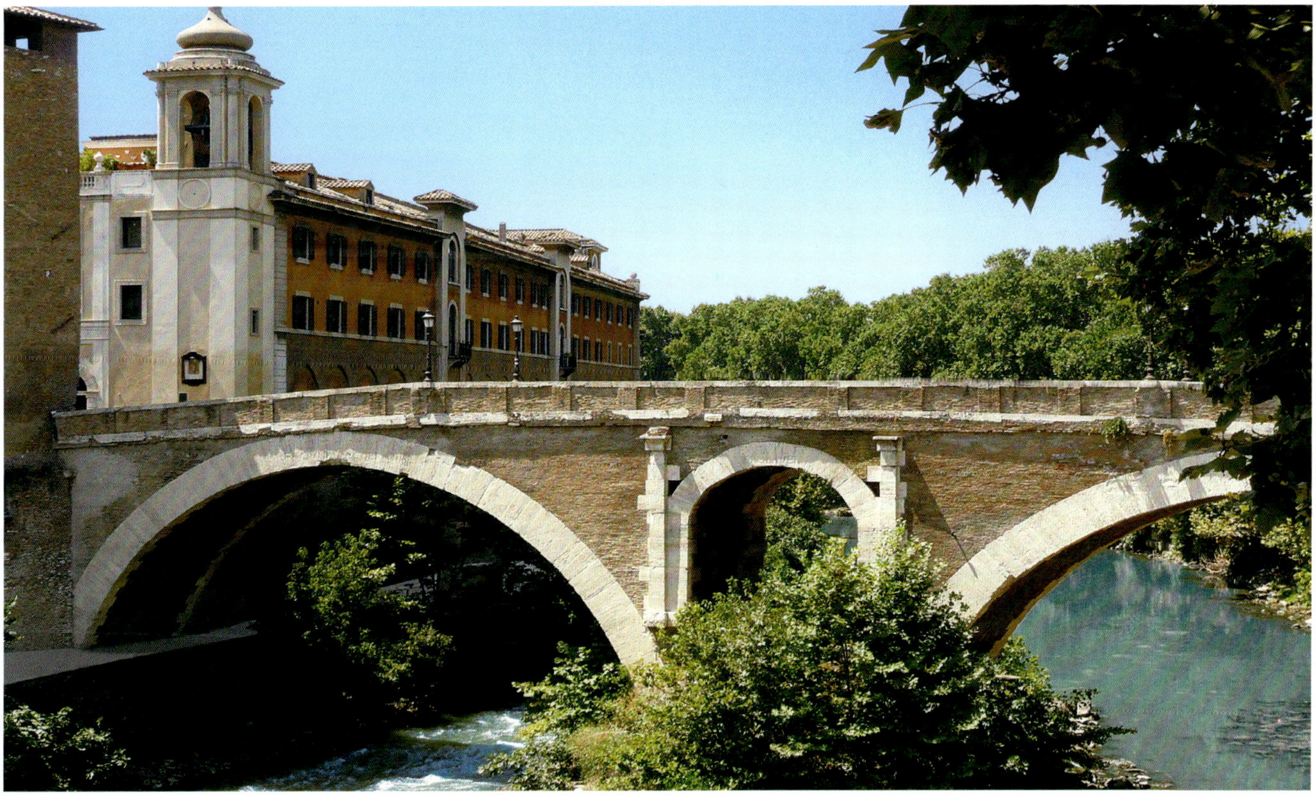

L FABRICIVS C F CVR
FACIVNDVM COERAVIT
LEPIDVS M F M LOLLIVS M F COS

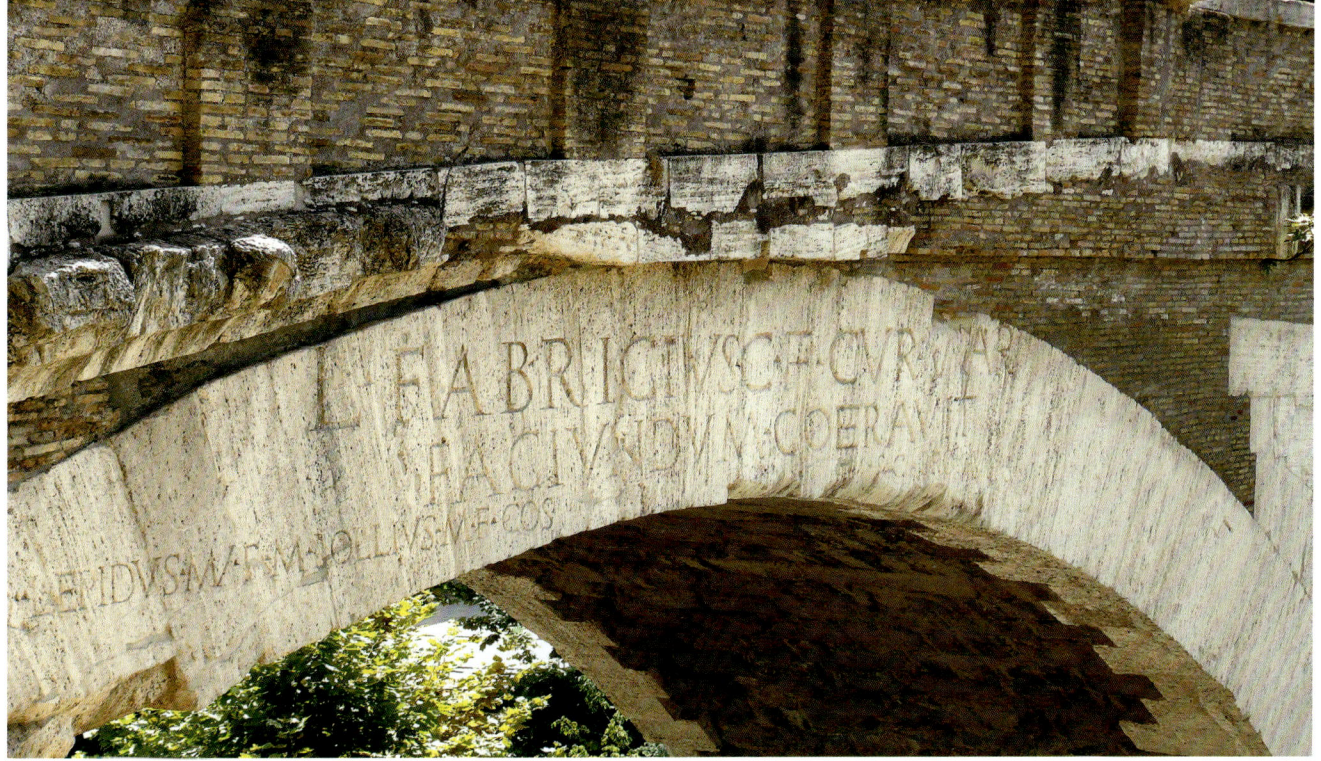

supported by a pier set in the middle of the river and rises toward the center over a small relieving arch. The bridge still bears on both sides its original inscriptions; those concerning its construction—*L. Fabricius C. f. curator viarum faciundum coeravit* ("Lucius Fabricius, son of Gaius, curator of roads, saw to [its] construction")—are engraved at the center of the arches, while those concerning the official acceptance—*Idemque probavit*—are at the center of the small axial arch. The Pons Cestius, dating to roughly the same period, was built on the other side of Tiber Island as a continuation of the Fabricius. About 48 meters in length, this bridge was completely transformed by restoration work at the end of the nineteenth century, but some of its original structure is known, including a large central arch and both its shoulders, which are pierced by smaller relieving arches. The stone used is travertine with blocks of peperino in the intradoses of the arches and white marble for the parapets.

Not surprisingly, many bridges were constructed in the imperial period, beginning with the Pons Agrippae, made by the Roman general Agrippa during the last quarter of the first century B.C. to join the Campus Martius with the Regio XIV Trans Tiberim, the area of the city now known as Trastevere. In the middle of the first century A.D., Nero built the Pons Neronianus—remains of which were still visible in the eighteenth century—immediately downstream from the modern Ponte Vittorio Emanuele. In A.D. 134 the emperor Hadrian completed another new bridge, the Pons Hadriani, or Aelius, creating a dramatic entry route to his mausoleum, today's Castel Sant'Angelo. This bridge measured about 150 meters long and 11 meters wide, and was composed of three central arches (each with a span of 18.4 meters), two smaller side spans, and a long eastern access ramp pierced by two relieving arches.

The closing centuries of ancient Rome saw the construction of two additional bridges. The first was perhaps ordered by the emperor Marcus Aurelius around A.D. 180, although some

The Ponte Cestio (ancient Pons Cestius), erected during the same period as the Pons Fabricius on the opposite side of the Tiber Island.

scholars argue that it was made by Caracalla between A.D. 211 and 217. The Pons Aurelius stood on the site of today's Ponte Sisto. Following restoration work executed by the emperor Valentinian in A.D. 365–67 (including construction of an honorary arch near the left shoulder of the bridge), the structure was renamed the Pons Valentiniani. The Pons Probi was erected by the emperor Probus between A.D. 276 and 282, a few hundred meters downstream from the Pons Aemilius and opposite the Aventine. Two letters of the statesman Quintus Aurelius Symmachus from the end of the fourth century A.D. report a curious anecdote concerning its reconstruction, when the bridge was renamed the Pons Theodosii: it seems the work site was directed by an architect named Auxentius, but he fled before completion of the bridge when he found himself the object of a series of pranks and sabotages.

VITRUVIUS AND THE FOUNDATIONS OF BRIDGES

The principal problem faced by the engineers of ancient Rome when constructing a stone bridge was the creation of underwater foundations. Although Roman foundations for bridges never approached those imagined by the eighteenth-century artist Giambattista Piranesi in his drawings, the building of solid supports on the ground was a priority for the stability of the entire structure. Consequently, engineers sought to make pylons as large as possible to balance the relatively reduced depth of the foundations, and they did so even when it meant reducing the span of the arches. The presence of water not only made digging foundations more difficult, but also impeded the normal procedure for setting lime mortar and thus interfered with the use of *opus caementicium*. However, the Romans managed to devise a solution, although a complex one.

In his detailed description of the construction of underwater foundations, Vitruvius (5.12.2) begins with this prescription for materials: "Masonry which is to be in the sea must be constructed in this way. Earth is to be brought from the district which runs from Cumae to the promontory of Minerva, and mixed, in the mortar, two parts to one of lime." He refers here to the pozzolana of the Phlegraean Fields required to create a mortar and mixed in a proportion of two parts of pozzolana to each part of lime, giving the mixture hydraulic properties. Even in the presence of water, the mortar would begin to harden after a period of about four days.

Vitruvius (5.12.3) then describes the building techniques for these foundations: "In the place marked out, cofferdams, formed of oak piles and tied together with chains, are to be let down into the water and firmly fixed. Next, the lower part between them under the water is to be leveled and cleared with a platform of small beams laid across, and the work is to be carried up with stones and mortar as above described, until the space for the structure between the dams is filled." The system was different in the absence of pozzolana, as Vitruvius explains (5.12.5): "Double cofferdams bound together with planks and chains are to be put in the place marked out. Between the supports, clay in hampers made of rushes is to be pressed down. When it is well pressed down and as closely as possible, the place marked out by the enclosure is to be emptied with water screws and waterwheels with drums, and so dried. Here the foundations are to be dug."

In conclusion, Vitruvius (5.12.5–6) provides instructions for the methods to adopt according to the nature of the terrain: "If the foundations are on the sea bottom, they are to be emptied and drained to a greater width than the wall to be built upon them, and then the work is to be filled in with concrete of stone, lime and sand. But if the bottom is soft the foundations are to be charred piles of alder and olive filled in with charcoal. . . . The wall is then raised of squared stones with joints as long as possible, so that the middle stones may be well tied together by the jointing. The inside of the wall is then to be filled in with rubble or masonry."

CUTS AND TUNNELS

The problems encountered when building a bridge or viaduct that was designed to connect two elevated sites separated by a wide gap (a riverbed or a cliff, for example) were hardly the only difficulties confronted by ancient engineers in their efforts to create roads that were passable for travelers. Sometimes they faced exactly the opposite situation—a natural barrier that blocked the route. So it was that the Romans, in part following a tradition of Etruscan origin, rather than elongating or altering the route of a road, set about cutting large openings through rock walls, creating what Pliny the Elder (*N.H.* 36.24) referred to as *viae per montes excisae*. One of the most spectacular examples of this technique is preserved along the Via Cassia, in the volcanic basin of Baccano, where a passage for a road was made by digging a trench extending 250 meters in length and reaching a depth of 25 meters. Similar stretches of road have been identified along the route between Castel d'Asso and Bomarzo.

These cuts in the rock usually resulted in a trench with nearly vertical or slightly arching walls that recalled the bottle-shaped walls of Etruscan origin. Alternatively, the flank of the hill was cut away, carving a roadway in the rock itself. This was the case, for example, with the Via delle Gallie, which ascends the valley of the Dora Baltea River toward the Little and Great Saint Bernard passes. In the Augustan period a road almost 5 meters wide was made here by cutting the rocky spur over a distance of 222 meters and up to a height of 12.75 meters; at the center of the cutting operation the rock was modeled not only into the milestone, with its indication of the distance to Augusta Praetoria (modern Aosta), but also a true "triumphal" arch, almost as though celebrating the accomplishment. Even more monumental was the cut of the Pesco Montano, made at Terracina during the reign of Trajan (A.D. 98–117). The operation was colossal, since to provide the Via Appia with access to the sea, the bedrock had to be cut away to create a wall more than 36 meters high. Incised on the surface, and still visible, is the numeral CXX, indicating the measurement of 120 Roman feet and relating to the height of the cut.

In other circumstances, the Romans found it necessary to dig tunnels of even greater length. Among the most monumental tunnels still in existence are those in Campania, including the Grotta di Cocceio and the famous Crypta Neapolitana, both made by the architect Lucius Cocceius Aucto during the Augustan period. With a length of 705 meters and a height of 4 to 5 meters, the Crypta Neapolitana connects Naples with Pozzuoli. Although the tunnel was ventilated and illuminated by shafts in its upper area that extended along the entire length, Seneca (*Epistles* 57) described the tunnel to his friend Lucilius with these dolorous words: "No place could be longer than that prison; nothing could be dimmer than those torches, which enabled us, not to see amid the darkness, but to see the darkness. But, even supposing that there was light in the place, the dust, which is an oppressive and disagreeable thing even in the open air, would destroy the light; how much worse the dust is there, where it rolls back upon itself, and, being shut in without ventilation, blows back in the faces of those who set it going!"

The Ponte Elio (ancient Pons Aelius) in an engraving by Giambattista Piranesi (mid-eighteenth century). Note the artist's imaginary reconstruction of the foundation system.

GLOSSARY

acroterion: figural decorative element, usually with a plinth, located at the lower corners or apex of a pediment.
adyton (Gr.), **adytum** (Lat.): literally, "inaccessible," a space in a Greek or Roman temple reserved for priests.
agora: central square or marketplace in a Greek city, which served as the center of economic and political activities.
ambulacrum: a covered walkway.
ambulatory: any place in which to process or walk.
amphitheater: Roman open-air performance space used for gladiatorial combats, mock sea battles, hunting spectacles, and other events.
analemma (pl. analemmata): support wall.
anathyrosis: contact band on the joining surfaces of a squared block.
anta (pl. antae): pillar terminating the end of a wall, often tapering and crowned by a capital.
antefix: ornamental piece (often with palmette decoration) located at the termination of roof tiles, on the eaves, or on the crest of a ridge.
Archaic period: one of the stylistic divisions of the history of ancient Greek art, usually dated to ca. 650–480 B.C. The other standard divisions are the Geometric, Classical, and Hellenistic.
architrave: rectangular beam or stone lintel placed in a horizontal position and usually supported on a pair of columns or piers.
archivolt: continuous architrave molding of the face of an arch.
ashlar: hewn or squared stone used in masonry to make horizontal courses.
avant-corps: part of a building projecting prominently from the main block.

base: lowest member of an architectural order, the support for a column, shaft, or pillar; also the support for an altar or statue.
bothros: pit dug in the earth, into which libations were poured or offerings thrown.
capital: topmost member of a column, pillar, or anta, located above the shaft.
caldarium: hot-room in a Roman bath.
cella: primary space of a temple and site of the cult image.
centering: temporary structure, usually wooden, used in the creation of an arch or a vault.
chora: agricultural area of a Greek city.
cipteral: having a double row of columns on each side, i.e., a double peristyle.
clamp: element, usually metal, used to connect two stone blocks.
Classical period: one of the stylistic divisions of the history of ancient Greek art, usually dated to ca. 500–323 B.C. (thus ending with the death of Alexander the Great). The other standard divisions are the Geometric, Archaic, and Hellenistic.
clast: sedimentary rock composed of fragments of older rock.
concrete bed: foundation method involving a layer of concrete that covers the entire area.
cornice: top, projecting section of an entablature; any molding that projects.
crepidoma: platform or base of a Greek temple, usually composed of three or four steps.
cyma: edge of a roof; terracotta or marble gutter with or without molding.
diaton: stone block equal in size to the width of a wall; also called a bond or bonding stone, through stone, or parpen.
distyle: having two columns.
domus: Roman home.
drachma: silver coin of ancient Greece; later the basic monetary unit of Greece.
drum: cylindrical block forming a column.
echinus: circular molding with a convex profile beneath the abacus in a Doric capital and beneath the volute in an Ionic capital.
elexion: stress at the curve of an element.
emplekton: in Greek architecture the core of aggregate materials filling the hollow between facing walls.
entablature: upper part of an order, composed of the architrave, frieze, and cornice.
entasis: slight convex curving on the shaft of a column.
epicranitis: decorative molding above the capitals of a row of pilasters or columns.
eschara: Greek hearth and site of sacrifice.
euthynteria: uppermost course of a Greek foundation.
extrados: outer convex face of an arch or vault.
formwork: usually temporary mold into which *opus caementicium* (concrete) is poured.
forum: Roman form of the Greek agora; central open space usually surrounded by buildings.
framework: bearing structure composed of the connection of various elements.
frieze: section between the architrave and the cornice of an entablature, usually decorated.

Geometric period: one of the stylistic divisions of the history of ancient Greek art, usually dated to ca. 900–700 B.C. The other standard divisions are the Archaic, Classical, and Hellenistic.
grooves: longitudinal fluting along the shafts of columns and pillars.
Hellenistic period: one of the stylistic divisions of the history of ancient Greek art, usually dated to ca. 323–31 B.C. (from the death of Alexander the Great to the battle of Actium, which established Augustus as emperor of Rome). The other standard divisions are the Geometric, Archaic, and Classical.
heroön: shrine or monument dedicated to a Greek or Roman hero.
hexastyle: having six columns.
hypocaust: underground chamber or duct of the Roman system of central heating, through which hot air circulated to heat the rooms above.
in antis: facade of a building with columns inserted between the antae of the side walls.
insula: city block in a Roman city, created by the crossing of streets.
intrados: inner curve of an arch or vault, also called the soffit.
megaron: main area of a Minoan and Mycenaean palace, the plan of which led to the simple shape of a temple and to that of the oikos.
metope: on a Doric frieze the smooth square space between two triglyphs, often decorated with figural reliefs.
monopteros, monopteral: having columns only, without a cella.
naos: similar to a cella, the primary chamber of a Greek temple.
nymphaeum: shrine dedicated to the Nymphs, female deities of nature, especially water.
odeum: small roofed building usually used as a theater and for the presentation of musicals and recitations.
oikos: Greek house or household; also a shrine with a single room.
opisthodomos: space behind the cella and open to the exterior.
opus: Latin for *work*, hence a construction technique.
opus caementicium: construction technique based on the use of aggregates mixed in mortar; concrete.
opus incertum: wall-facing for concrete, of irregularly shaped small stones.
opus isodomus: masonry of equally sized blocks set in regular rows
opus latericium: regular courses of lateral brick work.
opus listatum: alternating courses of brick and small blocks of stone.
opus mixtum: courses of bricks alternated with small stone work.
opus quadratum: squared stones laid in parallel courses.
opus reticulatum: concrete faced with small squared stones arranged diagonally.
opus testaceum: rubble and concrete faced with whole or broken bricks.
opus vittatum: wall-facing constructed with continuous horizontal courses.
order: column or row of columns in one of the standard styles: Doric, Tuscan, Ionic, Corinthian, or composite.
orthostat: stone in a masonry wall that has only one face showing.
palaestra (pl. palaestrae): exercise ground or public building for athletes.
palmette: decorative motif of leaves spread like a fan.
pediment: triangular structure or gable on top of the facade of a building with a pitched roof.
peribolos: enclosure within a sacred area; wall or colonnade surrounding such an area.
peristyle: having a colonnaded portico.
pilaster: square or rectangular shallow pier, projecting only slightly from a wall.
pisé: construction technique based on the use of an argillaceous mixture tightly packed into formworks.
plateia: large road in a Greek city.
plinth: square block serving as a base.
polis (pl. poleis): Greek city-state composed of the city inside its walls (*asty*) and the agricultural area (*chora*).
portico: porch; porticoed building.
pronaos: vestibule of a Greek temple leading to the cella.
propylon: entrance; free-standing gateway.
prostyle: having free-standing columns on the facade.
pulley: simple mechanism composed of a wheel and a cord, for lifting loads.
putlog holes: holes in a wall to support scaffolding during construction.
quoin: dressed stone at the corner of a building.
reins: in a vault, the parts between the crown and the spring or abutment
shaft: element of a column or pillar between the base and capital.
socle: lowest part of a wall or the base on which the wall stands.
stoa: porticoed building.
stylobate: level above the crepidoma on which the colonnade of a building stands.
temenos: sacred enclosure in which a Greek temple stands.
tepidarium: room with warm water for bathing.
tetrastyle: having a portico with four columns.
thalamos: circular chamber with a high vaulted or corbeled roof.
thesauros (pl. thesauroi): temple-shaped building for the preservation of ex-votos.
tholos (pl. tholoi): circular building with columns.
toichobate: level structure on which a wall stands.
torus: convex molding, especially at the base of a column.
trellis: structure composed of interwoven rushes or other plant elements.
triglyph: element of a Doric frieze composed of vertical grooves.
tristyle: having three columns.
tympanum: flat surface area inside a pediment.

BIBLIOGRAPHY

GENERAL WORKS AND ANCIENT SOURCES

Aulus Gellius. *The Attic Nights of Aulus Gellius*. 3 vols. Translated by John Carew Rolfe. New York, 1927.

Barletta, Barbara A. *The Origins of the Greek Architectural Orders*. New York, 2001.

Bozzoni, Corrado, et al. *L'architettura del mondo antico*. Rome, 2006.

Cassius Dio. *Dio's Roman History*. 9 vols. Translated by Earnest Cary and Herbert B. Foster. Cambridge, Mass., 1990.

Cato, Marcus Porcius. *On Agriculture*. Translated by William Davis Hooper. Cambridge, Mass., 1993.

Ciarallo, Annamaria, and Ernesto De Carolis, eds. *Homo faber: Natura, scienza e tecnica nell'antica Pompei*. Exh. cat. Museo Archeologico Nazionale, Naples, 1999.

Cicero. *Philippics*. Translated by Walter C. A. Ker. Cambridge, Mass., 1995.

Coarelli, Filippo. *Roma*. Milan, 1994.

"Comment construisaient les grecs et les romains." *Dossiers de l'archéologie* (November/December 1977).

The Digest of Justinian. Translated and edited by Alan Watson. Rev. ed. Philadelphia, 1998.

Diodorus Siculus. *Diodorus of Sicily*. Translated by C. H. Oldfather. Cambridge, Mass., 1989.

Frontinus, Sextus Julius. *Stratagems: Aqueducts of Rome*. Translated by Charles E. Bennett and Mary B. McElwain. Cambridge, Mass., 1997.

Ginouvés, René. *Dictionnaire méthodique de l'architecture grecque et romaine*. Vol. 1, *Matériaux, techniques de construction, techniques et formes du décor*. Vol. 2, *Eléments constructifs: Supports, couvertures, aménagements intérieurs*. Paris, 1985, 1992.

Greco, Emanuele. *Archeologia della Magna Grecia*. Rome, 2008.

———. *L'architecture romaine: Du début du IIIe siècle av. J.-C. à la fin du Haut-Empire*. 2 vols. Paris, 1996.

Gros, Pierre, ed. *Vitruvio, De architectura*. 2 vols. Turin, 1997.

Hellmann, Marie-Christine. *L'architecture grecque*. Vol. 1, *Les principes de la construction*. Vol. 2, *Architecture religieuse et funéraire*. Paris, 2002, 2006.

Herodotus. Translated by A. D. Godley. Cambridge, Mass., 1996.

Hesiod. *Homeric Hymns; Epic Cycle; Homerica*. Translated by Hugh G. Evelyn-White. Cambridge, Mass., 1998.

Landels, John G. *Engineering in the Ancient World*. Berkeley, 1978.

La Rocca, Eugenio, Mariette De Vos, and Arnold De Vos. *Pompei*. Milan, 2002.

Lauter, Hans. *Die Architektur des Hellenismus*. Darmstadt, 1986.

Lippolis, Enzo, Monica Livadiotti, and Giorgio Rocco. *Architettura greca: Storia e monumenti del mondo della polis dalle origini al V secolo*. Milan, 2007.

Livy. *History of Rome*. 14 vols. Translated by B. O. Foster et al. Cambridge, Mass., 1998.

Mazarakis Ainian, Alexander. *From Rulers' Dwellings to Temples: Architecture, Religion and Society in Early Iron Age Greece (1100–700 B.C.)*. Jonsered, 1997.

Mertens, Dieter. *Städte und Bauten der Westgriechen: Von der Kolonisationszeit bis zur Krise um 400 vor Christus*. Munich, 2006.

Orlandos, Anastasios K., and Ioannes N. Travlos. *Lexicon archaion architektonikon horon*. Athens, 1986.

Pausanias. *Description of Greece*. 5 vols. Translated by W.H.S. Jones, H. A. Ormerod, and R. E. Wycherley. Cambridge, Mass., 1992.

Pliny (the Elder). *Natural History*. 10 vols. Translated by H. Rackham et al. Cambridge, Mass., 1995.

Pliny (the Younger). *Letters and Panegyricus*. Translated by Betty Radice. Cambridge, Mass., 1989.

Plutarch. *Lives*. 10 vols. Translated by Bernadotte Perrin. Cambridge, Mass., 1993.

———. *Moralia*. 14 vols. Translated by Frank Cole Babbitt et al. Cambridge, Mass., 1989.

Seneca. *Epistles*. Translated by Richard M. Gummere. 3 vols. Cambridge, Mass., 1996.

Settis, Salvatore, ed. *Civiltà dei Romani: La città, il territorio, l'impero*. Milan, 1990.

———. *Civiltà dei Romani: Un linguaggio comune*. Milan, 1993.

Statius. *Silvae*. Translated by D. R. Shackleton Bailey. Cambridge, Mass., 2003.

Strabo. *Geography*. 8 vols. Translated by Horace Leonard Jones. Cambridge, Mass., 1988.

Suetonius. *Suetonius*. 2 vols. Edited and translated by J. C. Rolfe. Cambridge, Mass., 1998.

Tacitus. *The Annals*. 5 vols. Translated by John Jackson. Cambridge, Mass., 1998.

Torelli, Maria, and Theodorus Mavrojannis. *Grecia*. Milan, 1997.

Traina, Giusto. *La tecnica in Grecia e a Roma*. Rome, 1994.

Vitruvius. *On Architecture*. 2 vols. Translated by Frank Granger. Cambridge, Mass., 1998.

White, K. D. *Greek and Roman Technology*. Ithaca, 1984.

Wilson Jones, Mark. *Principles of Roman Architecture*. New Haven, 2000.

Wright, George R. H. *Ancient Building Technology*. Leiden, 2005.

NATURAL BUILDING MATERIALS: STONE AND MARBLE

Barresi, Paolo. *Province dell'Asia Minore: Costo dei marmi, architettura pubblica e committenza*. Rome, 2003.

Bessac, Jean-Claude, et al. *La construction en pierre*. Paris, 1999.

Caley, Early Radcliffe, and John F. C. Richards, trans. and eds. *Theophrastus on Stones*. Columbus, 1956.

De Nuccio, Marilda, and Lucrezia Ungaro, eds. *I marmi colorati della Roma imperiale*. Exh. cat. Mercati di Traiano, Rome, 2002–3.

Gnoli, Raniero. *Marmora romana*. Rome, 1988.

Gruben, Gottfried. "Naxos und Delos: Studien zur archaischen Architektur der Kykladen." *Jahrbuch des Deutschen archäologischen Instituts* (1997), pp. 261–416.

Lazzarini, Lorenzo, ed. *Pietre e marmi antichi: Natura, caratterizzazione, origine, storia d'uso, diffusione, collezionismo*. Padua, 2004.

———. *Poikiloi lithoi, versiculores maculae: I marmi colorati della Grecia antica; Storia, uso, diffusione, cave, geologia, caratterizzazione scientifica, archeo-metria, deterioramento*. Pisa, 2007.

Mottana, A., and M. Napolitano. "Il libro 'Sulle pietre' di Teofrasto." In *Rendiconti della Lincei: Scienze fisiche e naturali* (September 1997), pp. 151–234.

Pensabene, Patrizio. *Le vie del marmo: I blocchi di cava di Roma e di Ostia; Il fenomeno del marmo nella Roma antica*. Rome, 1995.

Rockwell, Peter. *The Art of Stoneworking: A Reference Guide*. New York, 1993.

CLAY AND TERRACOTTA

Auberson, P. "La reconstitution du Daphnéphoréion d'Erétrie." *Antike Kunst* (1974), pp. 60–68.

Caruso, Nino. *Ceramica viva: Manuale pratico delle tecniche di lavorazione antiche e moderne, dell'oriente e dell'occidente*. Milan, 1989.

Cuomo di Caprio, Ninina. *La ceramica in archeologia: Antiche tecniche di lavorazione e moderni metodi d'indagine*. Rome, 1985.

Guest-Papamanoli, Anne. "L'emploi de la brique crue dans le domaine égéen à l'époque néolithique et à l'âge du bronze." *Bulletin de correspondance hellénique* (1978), pp. 3–24.

Popham, Mervyn R., P. G. Calligas, and L. H. Sackett, eds. *Lefkandi, II: The Protogeometric Building at Toumba; Part 2. The Excavation, Architecture and Finds*. Athens, 1993.

Righini, V. "Materiali e tecniche da costruzione in età preromana e romana." In *Storia di Ravenna: 1. L'evo antico*, edited by Giancarlo Susini, pp. 257–96. Venice, 1990.

LIME, MORTAR, AND PLASTER

Blanc, Nicole. "Les stucateurs romains: Témoignages littéraires, épigraphiques et juridiques." *Mélanges de l'École française de Rome: Antiquité* (1983), pp. 859–907.

Demierre, Brigitte. "Les fours à chaux en Gréce." *Journal of Roman Archaeology* (2002), pp. 283–96.

Mannoni, Tiziano, and Enrico Giannichedda. *Archeologia della produzione*. Turin, 1996.

Mielsch, Harald. "Römische Stuckreliefs." *Römische Mitteilungen*. Heidelberg, 1975.

———. *Römische Wandmalerei*. Darmstadt, 2001.

La peinture murale romaine dans les provinces de l'empire: Journées d'étude de Paris (Paris, September 23–25, 1982). BAR International Series 165. Oxford, 1983.

Strong, Donald, and David Brown, eds. *Roman Crafts*. New York, 1976.

CONSTRUCTION TECHNIQUES IN THE GREEK WORLD

British School at Athens. *Perachora: The Sanctuaries of Hera Akraia and Limenia; Excavations of the British School of Archaeology at Athens*, 1930–33. Oxford, 1940.

Cambitoglou, Alexander, et al. *Zagora*, 1. Sidney, 1971.

———, et al. *Zagora, 2: Excavation of a Geometric Town on the Island of Andros*. 2 vols. Athens, 1988.

Cooper, Nancy Kelly. *The Development of Roof Revetment in the Peloponnese*. Jonsered, 1989.

Coulton, J. J. "Post Holes and Post Bases in Early Greek Architecture." *Mediterranean Archaeology* (1988), pp. 58–65.

Dinsmoor, William Bell. *The Architecture of Ancient Greece: An Account of Its Historic Development*. New York, 1975.

Fagerström, Kåre. *Greek Iron Age Architecture: Developments Through Changing Times*. Göteborg, 1988.

Hellmann, Marie-Christine. *Recherches sur le vocabulaire de l'architecture grecque, d'après les inscriptions de Délos*. Athens, 1992.

Hemans, F. P. "The Archaic Roof Tiles at Isthmia: A Re-examination." *Hesperia* (1989), pp. 251–66.

Maquettes architecturales de l'antiquité: Regards croisés (Proche-Orient, Egypte, Chypre, bassin egéen et Grèce du Néolithique à l'époque hellénistique). Actes du Colloque (Strasbourg, December 3–5, 1998). Paris, 2001.

Martin, Roland. *Manuel d'architecture grecque: 1. Matériaux et techniques.* Paris, 1965.

Orlandos, Anastasios K. *Les matériaux de construction et la technique architecturale des anciens grecs.* 2 vols. Translated by Vanna Hadjimichali and Krista Laumonier. Paris, 1966, 1968.

Roux, Georges. *L'architecture de l'Argolide aux IVe et IIIe siècles avant J.-C.* Paris, 1961.

Vallois, René. *L'architecture hellénique et hellénistique à Délos jusqu'à l'éviction des Déliens.* 2 vols. in 3 bks. Paris, 1944–78.

Williams, Charles K., and Nancy Bookidis, eds. *Corinth: The Centenary, 1896–1996.* Princeton, 2003.

CONSTRUCTION TECHNIQUES IN THE ROMAN WORLD

Adam, Jean-Pierre. *Roman Building: Materials and Techniques.* Translated by Anthony Mathews. Bloomington, 1994.

Durm, Joseph. *Die Baukunst des Etrusker; Die Baukunst der Römer.* Stuttgart, 1905.

Giuliani, Cairoli Fulvio. *L'edilizia nell'antichità.* Rome, 1998.

Lancaster, Lynne C. *Concrete Vaulted Construction in Imperial Rome: Innovations in Context.* Cambridge, 2005.

Lugli, Giuseppe. *La tecnica edilizia romana, con particolare riguardo a Roma e Lazio.* 2 vols. Rome, 1957.

Rea, R., H. J. Beste, and L. C. Lancaster. "Il cantiere del Colosseo." *Römische Mitteilungen* (2002), pp. 341–75.

Spanu, M. "L'uso delle anfore nelle volte romane e tardo-antiche: Distribuzione e modalità." In *Daidalos: Studi e ricerche del Dipartimento di Scienze del Mondo Antico 8* (Viterbo: Università degli Studi della Tuscia, 2007), pp. 185–223.

Taylor, Rabun. *Roman Builders: A Study in Architectural Process.* Cambridge, 2003.

Volpe, R. "Un antico giornale di cantiere nelle terme di Traiano." *Römische Mitteilungen* (2002), pp. 377–94.

ENGINEERING AND TECHNIQUES AT THE WORK SITE

Adam, Jean-Pierre. "A propos du trilithon de Baalbek: Le transport et la mise en oeuvre des mégalithes." *Syria* (1977), pp. 31–63.

———. "Groma et chorobate: Exercices de topographie antique." *Mélanges de l'École française de Rome: Antiquité* (1982), pp. 1003–29.

Varène, P., and Jean-Pierre Adam. "Une peinture romaine représentant une scène de chantier." *Revue archéologique* (1980), pp. 213–38.

Zimmer, Gerhard. *Römische Berufsdarstellungen.* Berlin, 1982.

ANCIENT HYDRAULICS: BETWEEN TECHNOLOGY AND ENGINEERING

Ashby, Thomas. *The Aqueducts of Ancient Rome.* Oxford, 1935.

Bodon, Giulo, Italo Riera, and Paola Zanovello. *Utilitas necessaria: Sistemi idraulici nell'Italia romana.* Milan, 1994.

Bonnin, Jacques. *L'eau dans l'antiquité: L'hydraulique avant notre ère.* Paris, 1984.

Ginouvès, René. *L'établissement thermal de Gortys d'Arcadie.* Paris, 1959.

Glaser, F. *Antike Brunnenbauten ([krēnai]) in Griechenland.* Vienna, 1983.

de Haan, Nathalie, ed. *Cura aquarum in Campania.* Proceedings of the Ninth International Congress on the History of Water Management and Hydraulic Engineering in the Mediterranean Region (Pompeii, October 1–8, 1994). Leiden, 1996.

Tölle-Kastenbein, Renate. *Antike Wasserkultur.* Munich, 1990.

Werner, Dietrich. *Wasser für das antike Rom.* Berlin, 1986.

HEATING SYSTEMS AND BATHS

DeLaine, J., and D. E. Johnston, eds. *Roman Baths and Bathing.* Proceedings of the First International Conference on Roman Baths (Bath, March 30–April 4, 1992). Portsmouth, R.I., 1999.

Ginouvès, René. *Balaneutikè: Recherches sur le bain dans l'antiquité grecque.* Paris, 1962.

Hoffmann, Michela. *Griechische Bäder.* Munich, 1999.

Nielsen, Inge. *Thermae et balnea: The Architecture and Cultural History of Roman Public Baths.* 2 vols. Aarhus, 1990.

Les thermes romains: Table ronde; Selected Papers; Notes from a Meeting. École Française de Rome, November 11–12, 1988. Rome, 1991.

Yegül, Fikret. *Baths and Bathing in Classical Antiquity.* New York, 1992.

Wikander, Örjan, ed. *Handbook of Ancient Water Technology.* Boston, 2000.

ROADS, BRIDGES, AND TUNNELS

Busana, Maria Stella, ed. *Via per montes excisa: Strade in galleria e passaggi sotterranei nell'Italia romana.* Rome, 1997.

Catalano, Romilda. *Intus in tenebris: Scienza e tecnica nelle opere ipogee romane.* Naples, 2007.

Chevallier, Raymond. *Les voies romaines.* Paris, 1997.

Galliazzo, Vittorio. *I Ponti romani: 1. Esperienze preromane, storia, analisi architettonica e tipologica, ornamenti, rapporti con l'urbanistica, significato; 2. Catalogo generale.* Treviso, 1995.

Greco, Emanuele. "Nomi di strade nelle città greche." In *Koina: Miscellanea di studi archeologici in onore di Piero Orlandini*, edited by Marina Castoldi, pp. 223–29. Milan, 1999.

Quilici, Lorenzo, and Stefania Quilici Gigli. *Tecnica stradale romana.* Rome, 1992.

INDEX

Note: Page numbers in *italics* refer to illustrations.

Africanus, Publius Scipio, 202
Agrigento, 96, *96*, 201
Agrippa, 165, 166, 174, 178, 183, 205
Ahenobarbus, Domitius, 196
Aigosthena, *97*
Alexander the Great, 200
Ampurias, 47
Ancus Martius, 202
Antonius Pius, 40
apodyteria, 179–80
Apollodorus of Damascus, 128, 137, 138, 184
aqueducts, 159–66; centering arches of, 148–50; reservoirs for, 168
Aquileia, 15, *15*
arches: centering of, 148–52; design of, *132*, 132–35; Greek, 132, 148–52; in *opus caementicium*, 134–37, 150; in *opus quadratum*, 134–35; origins of, 131–32; Roman, 131–37, 148–51
architecture, origins of, 41–42
architraves, 135, 143–44, *144*
archivolts, *132*, 137
argillaceous mixture, *45*, 46, 47–48
Argos, *81*, 81–82, *83*, 98, 153
Aristotle, 166
Athens: quarries in, 9, 16; water in, 155, *155*, *158*, 159, 170, 172
Athens, Acropolis in: Erechtheum of, *100–101*, 110; Parthenon of, *142–43*; Propylaea of, *95*, 96; stone used at, 9; transportation of stone for, 140–41; work-site organization at, 108–9
Aucto, Lucius Cocceius, 207
Augusta Praetoria, 16
Augustus, 39–40, 126, 165, 166
Aulus Gellius, 111
Auxentius, 206

Babylonia, 199–200

Bassae, 9
baths, 175–86; Greek, 175–77; reservoirs for, 168; Roman, 137–38, 177–86; waterproofing for, 72, 76
Biancone, 15
block and tackle, 144–46, *146*
Botticino, 15
Brauron, 200
bricks, fired, 55–58; Greek use of, 55–56, 59; production of, 56, 56–58, *58*; Roman use of, 56, 127–30; size and shape of, *58*, 59, *60*
bricks, mud, 47–58; Greek use of, 50–55; production of, 47–49; Roman use of, 51, 127–28
bridges, 198–207
Brixia, 15
Byzes, 18

Caecus, Appius Claudius, 164, 196
caldaria, 179, 180, *180*
canals: stone, *104*, 104–5; water, 189
Capua, *126*, *138*, 147
Caracalla, 184, *185*, 206
Cassius, Dio, 137, 138, 174
castellum aquae, 168–69, *169*
Cato the Elder, 65–66, 116
Catulus, L. Cornelius, 134
centering, *148*, 148–52
Cetius Faventinus, Marcus, 76
Chersiphron, 143, *144*
chisels, 38, *39*
Cicero, 196
cisterns, 72, 166–68
clamps, 106–8, *107*
Claudius, 40, 166
clay, 42–59; extraction of, 44; in *formacei* walls, 46–47, *47*; Greek use of, 50–56, 59; location of deposits, 42–44; models made of, 81–82; origins of use, 42; in pipes, 60, *60*, 169; in *pisé*, 44–47; properties of, 42; quarries of, 42–44; Roman use of, 51,

56, 59–60; in tiles, 60, *60*, 82–90, 93; in trellis technique, 45, 45–46; types of, 44; wood combined with, 45–46, 51–55; working with, 44. *See also* bricks; terracotta
Cocles, Horatius, 202
columns: stone, 36–38, *37*; wood, 93–94
concrete. *See opus caementicium*
construction techniques. *See specific locations, materials, and techniques*
Copia, 76
Corinth: Acrocorinth at, *43*, 44; clay in, 42–44, *44*; Fountain of Glauke at, 8; roofs of, 85, *88*; springs at, 156–57, *157*; Temple of Apollo at, 9, *84*, 84–85, *85*
Corinth, Isthmus of, *187*, 187–90, *188*
Cosa, *111*, 112
cranes, 146, *147*
crepidomas, 96
Crete, baths of, 175, *175*. *See also specific sites*
Crypta Neapolitana, 207
Cusa, 10, *34*
Cyclopean walls, 97
Cyprus, 84

Darius, 200
Decius, 185
Delos: cisterns in, *168*; House of the Trident on, *75*; Sanctuary of Apollo on, 20–23, *22*, *23*, 94, *103*; walls in, 74, *75*, 98, *103*
Delphi: baths in, 175–76; Sanctuary of Athena Pronaia at, 73–74, *94*, *97*; Temple of Apollo at, 16–18, 31, 54, 87; transportation of stone to, 139–40, *141*
Demetrius I, 189
Democritus of Abdera, 131–32
Dentatus, Manius Curius, 164
diatons, 99, 101, 102, 112
Didyma, 38, *108*, *109*, 110, *140*

Digest of Justinian, 196, 198
Diocletian, 25, 28, *30*, 185
Diodorus Siculus, 39, 192
Diolkos, *188*, *189*, 189–90
direct-percussion tools, 38, *39*
Divordurum, 148–50
domes, 150–52
Domitian, 166

Egyptian diorite, 28
Eleusis, *49*, 141, 159, *159*
Eleutherai, 98, *98*
Ephesus: Celsus Library at, *35*; fountains of, *171*; gate of Tetragonos Agora at, 114, *114*; quarries in, *32*; Temple of Artemis at, 143–44, *144*
Epidaurus, 109
Eretria, 52–55, *53*, *54*, 200
Etruria, 15
Etruscans, 7
Euboea, *51*, 51–52, *52*
Eupalinos, 160, *160*, *162*, 162–63, *163*
euthynterias, 95–96
Evans, Arthur, 77
Exodus, 48

Fabricius, Lucius, 202, *203*, 205
Fidenae, 12–13, 118
Flamminius, Gaius, 196
forceps, *104*, 105–6, *106*
Forma Urbis Romae, 138, 178, 183
formacei walls, 46–47, *47*
foundations, of bridges, 206; of buildings: Greek, 92, *94*, 94–96, *96*; Roman, 116, 117–22
friezes, *89*, 90–91
frigidaria, 179, *179*, 180
Frontinus, Sextus Julius, 163–64, 166

Gortys, *94*, 176–77, *177*
gradines, 38
Greek construction techniques, 77–110. *See also specific locations, materials, and techniques*
Greek Islands, marble from, 16–23. *See also specific islands*
Grumentum, 182
gymnasiums, 175–76
gypsum, 61, 64

Hadrian, 137, 188, 205
hammers, 38, *39*
heating systems, 60, 176–77, 180–82, *181*
Herodes Atticus, 157
Herodotus, 16–18, 162–63, 199–200
heroön, 52
Hesiod, 76
Hierapolis, *173*
Hippodamus of Miletus, 191, *191*
hoisting methods, 102–6, *104*, 143, 144–48
hypocausts, 181–82, *182*

Iaitas, 72
indirect-percussion tools, 38, *39*
Iraq, Temple of Tell al-Rimah in, 132
isodomus walls, 99
Istanbul, 168, *169*
Isthmia, 85–87, *87*

Karphi, 77–79, *78*
keystones, 131, 132, *132*, 135, 150
kilns, 55–59, *56*, *65*, 65–66
Korres, Manolis, 140

Laconian roofs, 88, *88*, 93
large blocks, *92*, 92–93, 111–14
lavatories, 170–72
lead pipes, 169, *170*
Lefkandi, *51*, 51–52, *52*
Leptis Magna, 31, 40
Lerna, 82
Lesbian-style masonry, *97*, 98
levers, 106, *106*
lewises, *104*, 105
lime, 61–71
limestone: in Greece, 8–10, 84–86, *86*; in lime production, 61–62, 64–66; in Rome, 14–16
Liternum, *130*
Livy, 7, 112, 134, 172–74, 194, 198, 202
Longinus, Cassius, 196
Lupiae, 14, *15*

mallets, 38, *39*
marble(s): baking of, 64; cutting of, 18; Greek use of, 16–23; origins of use, 16, 20, 25; prices of, 25, 28, *30*; Roman use of, 24–30; sources of, 16–30; transportation of, 25, 40, 139–44; types of, 16–18, 25. *See also specific sites*
marble quarries: in Africa, 28–30; in Asia Minor, 26–28; excavation methods at, 30–38; formation of marble at, 16; in Greece, 16–23, 25–26; in Rome, 24–25; rough shaping at, 38
Marcus Aurelius, 205–6
Martial, 172
Maxentius, 131
Meandrius, 163
measurement systems, 110, *110*, 169
Meligalas, *199*
Mesopotamia, 132
Messina, 200–201
Metagenes, 143–44, *144*
Metapontum, *190*, 190–91
mining, 30–38, 44
Mnesicles, *17*, 96
models, architectural, 81–82, *83*, 110
mortar(s), 61–76; ingredients of, 61–71; in *opus caementicium* walls, 114–17; *opus signinum*, 71–73; for plaster and stucco, 73–75; preparation of, *70*, 70–71; special, 76; waterproofing with, 72–73, 76
Mummius, Lucius, 202
Mycenae, 6, 7, 154, *155*

Nemea, 176, *176*
Nero, 166, 178, 183, 184, 189, 205
Nichoria, 46, *46*
Nicopolis, 125, *126*, *130*, 165
Nitocris, 199–200
nymphaea, 170

Olympia: Heraion at, *90*, *91*, 91–94, *92*; Temple of Zeus at, 9, *10*
Olympic games, *90*, 91
open-pit quarries, *32*, 33, 44
opus caementicium, 114–22; in arches, 134–37; in foundations, 116, 117–22, *118*, *119*; preparation of, 114–17; in vaults, 136–37, 150–52; in walls, 59, 114–17, *116*, *117*, 123–25
opus incertum, 124, 124–25, *130*
opus latericium, 127
opus listatum, *126*
opus mixtum, 130–31

opus quadratum, 112–14, 123, *123*, 134–35
opus reticulatum, 124–27, *125*, *126*, 130–31
opus signinum mortar, 71–73
opus testaceum, 128–30
opus vittatum, 127, 131
opus vittatum mixtum, 131
Orata, L. Sergius, 180–81
orthostats, 101, 102, 112
Ostia, *71*, *76*, *178*
overburden, 33

Paestum, *124*, *125*
Palladio, Andrea, 76, 183, 184
Paradise Quarry, 10
Paros, marble from, 16–18, *19*, 34
Pausanias, 7, 9, 18, 91, 93, 177, 187–88
Peculiaris, Lucceius, *145*, 147
Peisistratus, 170
Pentelic marble, 16, *17*, 140–41
Perachora, 81–82, *83*, 87
Periander, 189
Pericles, 108
Persia, *199*, 200
picks, 38
pins, 106–8, *107*
pipes, 60, *60*, 76, 169, *170*
Piranesi, Giambattista, 206, *207*
pisé, 44–47
Pixodarus, 32
plaster, 73–75
Plato, 176
Pliny the Elder: on aqueducts, 165; on baths, 178; on bricks, 50; on clay construction, 46–47; on gypsum, 64; on kilns, 56–58; on lime, 62, 64, 66, 71; on marble, 16, 18; on mortar preparation, 70, 71; *Natural History*, 84; on roads, 188–89, 207; on sand in mortar, 69; on serpentine, 26; on sewage systems, 172; on special mortars, 76; on terracotta tiles, 82–84; on transportation from quarries, 32
Pliny the Younger, 118–19
Plutarch, 87, 88, 108–9
polygonal masonry, 97–98, 112
Pompeii: basilica at, 127, *128*; *castellum aquae* at, 168–69, *169*; House of Euthycus at, *73*; House of the Iliac Chapel at, *70*, 71; Porta Nuceria at, *73*, *124*; Stabian Baths at, 180, *180*; types of stone at, 14
Poseidonia, 9–10
Posidonius, 132
pottery shards, 71–73
Priene, 75
Probus, 206
Proto-Corinthian style, 85, 86–87
Ptolemy II Philadelphus, 201
pulleys, 144–48
punches (tool), 38, *39*

quarries, clay, 42–44. *See also* marble quarries; stone quarries
quicklime, 62

rainwater collection, 166–68
reservoirs, 166–68
retaining walls, 101
Rex, Quintus Marcius, 165
Rhamnus, *103*
roads: Greek, 140–41, 187–92; Roman, 13, 192–98, *197*, 207. *See also* bridges
Roman construction techniques, 111–38. *See also specific locations, materials, and techniques*
Rome (city): Aqua Marcia in, 148, *149*, 150, 165; aqueducts of, 163–66, *166*; Arch of Constantine in, *134*, 137; Basilica of Maxentius in, *136*, 137, *151*; Basilica of Neptune in, *24*; baths of, 177–80, 183–86, *184*, *185*, *186*; bridges of, *202*, 202–6, *203*, *204–5*; Campus Martius in, 183, *183*; Capitoline Hill in, *11*, 111, 112, *112*, 123; Cloaca Maxima of, 172–74; Flavian Amphitheater of, *120*, 120–22, *121*, *122*; Forum of, *29*, 134; Forum of Augustus in, 113, *115*; Forum of Caesar in, 113, *113*; fountains of, 170; lavatories of, 170–72; marble transported to, 25, 40; Mausoleum of Augustus in, *125*; Mausoleum of Cecilia Metella in, *61*; Palatine Hill in, 46, 62–63, 116, *116*; Pantheon of, *27*; Porta Maggiore in, *133*; Porticus Aemilia in, 116–17, *125*, *135*, 137; roads in, 192–94; Servian Wall around, 112; springs of, 156; stone quarries in, 12; Tabularium of, *67*, 134–35; Temple of Venus in, *118*, 119; Theater of Marcellus in, *12*, 13, *131*; Trajan's baths in, 137–38, *184*, 184–85; Trajan's Market in, *57*, 68–69, *129*; Tullianum of, 156, *156*; walls of, 111–13, 116–17, 126
roofs: of architectural models, 82; stone in, 78–79, 80–81; terracotta tiles on, 82–90, 93; wood in, 52, 54, 55, 78–79, 80, 86
Rufo, Titus Annius, 196

Salamis, 110, *110*
Samos, *50*, 160, *160*, 162, 162–63, *163*
Samothrace, *200*, 201
sand, 67–69, 70
saunas, 180
scaffolding, 23, *152*, 152–53
schist, 80–81
Scironian Road, 187
Scopas, 177
Segesta, 104, *105*
Selinunte, 201, *201*
Selinus, 36–37, 144
Seneca, 131–34, 172, 178, 207
Sennacherib, 159
Severus, Alexander, 166, 183
Severus, Septimius, 31, 138
sewage systems, 172–74
ships, 139, 141, 189, 189–90
Sicily, 10
Skyros, 25
slaked lime, 62–64, 66–67
socles, 52, 84
Sparta, 88
Spilia, *141*
Spina, 46
springs, 155, 155–57, *156*, 157
square masonry: Greek, 97–102, *102*; Roman, 112–14
Stabiae, 146
statio marmorum, 40
Statius, 197
stepped quarries, 32, *33*, 44
Stertinius, C. Lucius, 134
stone(s): baking of, 64–66; for bridges, 200–201; connections between, 106–8; for foundations, 94–95, 119; Greek use of, 7–10; lifting and placement of, 102–8, 143, 144–52; location

of, 8–10, 12–16; origins of use, 7; rise in use, 82, 84; Roman use of, 10–16; transportation of, 139–44; types of, 8–13; for walls, 97–102, 111–17; weathering of, 10, 14. *See also specific sites*
stone quarries, 30–40; discovery of, 32; excavation methods at, 30–38; in Greece, 8–10; organization of, 39–40; roads to, 188; in Rome, 12–16; rough shaping at, 38; specialization at, 31, 39; transportation from, 25, 32, 40, 139–44
stonecutters, 31, 39, 109–10
Strabo, 16, 24, 25, 189, 194, 195
straw, 48, 56
stucco, 73–75
substructures, 153
Suetonius, 25, 39
Sybaris, 191, *191*
Symmachus, Quintus Aurelius, 206
Syracuse, 10, *161*

Tacitus, 118
Tarquinia, 7
Tarquinii, 13–14
Tarquinius Priscus, 172
Tarquinius Superbus, 174
Tarraco, *126*
temples, terracotta on, 89–91. *See also specific sites*
tepidaria, 179, 180
terracotta, 59–60; in architectural models, 81–82; in pipes, 169; Roman use of, 59–60; in roof tiles, 82–90, 93; in temple decoration, 89–91
Thasos, *31*, *33*, 139, *139*, 140
Thebes, 48
Theophrastus, 61, 62–64, 93
Thermon, 89, 89–91
Thrace, *48*
Thucydides, 189
Thurii, *191*, 191–92
Tiberius, 25, 39, 56
tiles, 18, 60, *60*, 82–90, 93
Titus, 184
torchis, 45
Trajan, 137, 166, *184*
transportation: of ships across land, *189*, 189–90; of stone, 25, 32, 40, 139–44. *See also* roads

trapezoidal masonry, 98, *98*, 112
trellis construction, 45, 45–46
Turin, 16
Tychicus, C. Haterius, 147

Valentinian, 206
vaults: *vs.* arches, 132; centering of, 148–52; design of, 136, *136*; Greek, 148–52; in *opus caementicium*, 136–37, 150–52; Roman, 136–37, 148–52
Veii, 7, 111
Verona, 15
Vesuvius, Mount, 70
Via Appia, *123*, 123–24, *195*, 196, *196*, 198, 207
Via Cassia, 207
Via delle Gallie, 207
Via Mansuerisca, *197*
Via Sacra, *193*, 194
viaducts, 198–99
Vitruvius: on aqueducts, 161, 162, *162*; on architects, 4; on baths, 179, 181, 182, 184; on brick walls, 127–28; on clay construction, 45–46, 48–49; *De Re Edificatoria*, 4, *41*; on discovery of quarries, 32; on foundations, 118, 206; on hoisting machines, 106, 145–48, *147*; on lime production, 62, 66; on mortar preparation, 70; on mortar use, 72, 73; on mud bricks, 50, 51, 127–28; on *opus caementicium*, 114–16, 117; on *opus incertum*, 124; on *opus reticulatum*, 124; on origins of architecture, 41–42; on pipes, 169; on plastering, 74; on pozzolana, 69; on rainwater, 166; on roads, 197; on sand in mortar, 67–69; on springs, 156; on substructures, 153; on transportation of stone, 143–44; on types of stone, 10, 12, 13–14; on walls, 97, 99, 101, 127–28; on water, need for, 155; on wells, 158
voussoirs, 132
Vulca, 111

walls: in architectural models, 82; brick, 50, 52, 127–30; combination of types, 130–31; framed, 46–47, *47*; Greek, 97–102; *opus caementicium*, 59, 114–17, 123–25; *opus quadratum*, 112–14, 123; plaster on, 73–75; Roman, 111–17, 123–31; stone, 7. *See also specific sites*
water, 155–74; bridge foundations under, 206; in cisterns, 166–68; clay mixed with, 44, 45; distribution of, 168–69; in fountains, 170; in lavatories, 170–72; in lime production, 62–64; in mortar, 67; in nymphaea, 170; in reservoirs, 166–68; in sewage systems, 172–74; from springs, 155–57; from wells, 157–59. *See also* aqueducts; baths
waterproofing, 72–73, 76
weathering of stone, 10, 14
wedges, 32, *33*, 34–36
wells, 34–36, 157–59
windlasses, 144–46
wood: in bridges, 200, 202; clay combined with, 45, 45–46, 51–55; in columns, 93–94; in foundations, 92, 119–20; in roofs, 52, *54*, 55, 78–79, 80, 86; transition to stone from, 84–85. *See also specific sites*
work sites, 139–53; centering of arches and vaults at, 148–52; connecting of stones at, 106–8; dressing of stones at, 102; lifting of stones at, 102–6, 143, 144–58; models and drawings at, 110; organization of, 108–10, 138; placement of stones at, 106–8, 148–52; reports on, 109; scaffolding at, 23, 152–53; substructures at, 153; transportation from quarry to, 139–44

Xanthos, *93*
Xerxes, *199*

Yria, 18–20, *20*, *21*

Zagora, *79*, 79–81, *80*
Zakros, *175*